Interpolation and Extrapolation Optimal Designs 1

Series Editor
Nikolaos Limnios

Interpolation and Extrapolation Optimal Designs 1

Polynomial Regression and Approximation Theory

Giorgio Celant
Michel Broniatowski

iSTE

WILEY

First published 2016 in Great Britain and the United States by ISTE Ltd and John Wiley & Sons, Inc.

ISTE Ltd
27-37 St George's Road
London SW19 4EU
UK

www.iste.co.uk

John Wiley & Sons, Inc.
111 River Street
Hoboken, NJ 07030
USA

www.wiley.com

Library of Congress Control Number: 2016933881

British Library Cataloguing-in-Publication Data
A CIP record for this book is available from the British Library
ISBN 978-1-84821-995-3

Contents

Preface

This book is the first of a series of three which cover a part of the field called optimal designs, in the context of interpolation and extrapolation. This area has been studied extensively, due to its numerous applications in engineering, physics, chemistry, and more generally, in all fields where experiments can be planned according to some expected accuracy on the resulting conclusions, under operational constraints.

The context of the present volume, which has been considered, is very specific by choice. Indeed reducing the model to the case where the observations are real numbers, hence unidimensional, and the environmental variable being unidimensional too, most of the concepts gain in clarity; also the tools leading to optimal solutions are quite accessible, although they require some technicalities. This choice will help the reader to consider more general models, keeping in mind the basic ingredients which are developed in more involved situations, at the cost of some additional regularity assumptions. This is a well-known way to proceed in science.

The focus of this book is statistics. The contents of the book are mostly real analysis and approximation of functions. This duality in the arguments is not surprising, and is a constant in most advanced fields in statistics, and also in various other disciplines. It happens, and this is a major fact in the present field of statistics, that optimal designs are obtained as special problems in the theory of functional approximation. So, those two fields, namely statistics and numerical analysis, meet in the present setting.

The framework of the present book is thus one of the classical real analysis, of the basic tools in algebra together with standard basic tools in statistics. The reader may also be interested in the chronological aspect (or historical aspect) of the development, as the authors have been. Although of statistical concern, the main arguments used in this book stem from the theory of the uniform approximation of functions. This argument has a long and interesting history, from the pioneering works of Lagrange and Legendre, followed by the contributions of Chebyshev and Markov, continuing through famous results by Lebesgue; the results obtained by Bernstein and Vitali provide sharp and interesting insights into the properties of polynomials, and are of interest in the accuracy of the approximation of functions. Borel provided a final description to the approximation results due to Chebyshev. Erdös gave a strong improvement to Bernstein's contribution in the rates of approximation. The reader will find those elements in the core of the present volume and in the Appendices.

Most optimal designs do not result as analytic solutions for approximation problems. Algorithmic solutions have been developed over the years: the optimal designs are obtained through algorithms which were developed in the field of the theory of the uniform approximation of functions, and are nowadays, important tools in numerical analysis. Henceforth, those algorithms have also been studied by statisticians; such is the case for the Remez algorithm, and to its extension by de Boor and Rice to constrained cases. This volume presents these tools in the statistical context.

The choice of the statistical context is restricted to the regular one, in the sense that all random variables which describe the variability of the inputs are supposed to be independent, essentially with the same distribution, with finite variance, hence allowing the least mean square paradigm in the field of linear models. This is the basic framework. The companion volumes consider nonlinear models, heteroscedastic models, models with dependence in the errors, etc.

This book results from our teaching, both in the University of Padova and in University Pierre and Marie Curie (Sorbonne University) in Paris. This corresponds to a one semester course in statistics or in applied mathematics.

The authors express their gratitude to their families and friends, who provided inestimable support during the completion of this work.

Giorgio CELANT, Padova
Michel BRONIATOWSKI, Paris
February 2016

Introduction

I.1. The scope of this book

This book is devoted to the construction of optimal designs for the estimation of a regression function, either in interpolation or in extrapolation. This deserves some definition.

We assume the following model to hold. A random variable $Y(x)$ is observed whose expectation depends on some environment x, through a function f,

$$E(Y(x)) = f(x).$$

The random measurement $Y(x)$ differs from its expectation by an additive random error term, say ε, with zero mean. We assume that the distribution of ε may depend on the environment x and that ε has a finite variance; in this book, this variance is assumed to be constant with respect to x. In our context, little is known by the experimenter: the function f is unknown, as is the distribution of the random error ε. However, in this volume the function f will be assumed to belong to a known class of functions, so that no misspecification issues can occur. This choice is, therefore, not a matter of convenience, but a basic assumption.

The experimenter is able to perform measurements of $Y(x)$ under any environment x in some range. For some reason, the number of such measurements is fixed, say n. The experimenter will choose a number g, of experimental conditions, say $x_0, .., x_{g-1}$ and, for any of those, he will decide to perform a number of measurements of the variable of interest $Y(.)$ at those locations, say n_j measurements in the environment x_j. A design is defined by the collection of environmental points $x_j^{\prime}$s and the number of replications $n_j^{\prime}$s, which sum to n, together with the value of g.

The aim of the experimenter is to estimate the value of f at some x. Here are some different situations. In the first case, Y can be measured under condition x. The experimenter will perform all measurements under x and estimate $f(x)$ by the empirical mean of the measurements. In the second case, an experiment under x is not feasible; however, it can be performed at any x' in the neighborhood of x. The value $f(x)$ will be estimated by a simple combination of the measurements of the Y's performed on a design defined on this neighborhood. The design should be such that the mean square error (MSE) of the estimate of $f(x)$ should be minimal. We have introduced a natural criterion for optimality.

Accelerated runs, or prediction problems, describe a further situation. The environment x lies outside the range of possible measurements: there exists a neighborhood of x on which no measurement can be performed. The design will be built on the entire observable domain; the estimate of $f(x)$ will also, in this case, be a simple function of the observations on the design, and the optimality criterion leading to its definition will still be the MSE.

Sometimes, the experimenter does not choose the value of x beforehand. He will build the design in a way such that the resulting estimator of $f(x)$ bears some property independent of the exact value of x. This introduces designs which are optimal uniformly upon the MSE of the estimate of $f(x)$, when x belongs to some set.

We emphasize that the concern of these problems lays in the construction of the design, and not on the precise form of the estimate, which will be of a basic and quite natural nature in this context, namely Gauss Markov estimators.

We now put forward some notation.

The environments belong to some interval in $\mathbb{R}$, say $[-1, 1]$ for convenience. Therefore, the setting is very specific with respect to the usual schemes in regression analysis, where the regressors are defined in spaces with many dimensions. The function f will also satisfy strong and simple hypotheses. A prototype is when f is a polynomial defined on $\mathbb{R}$ with known degree $g - 1$. A natural extension is to assume that f is in the span of a family of g independent real valued functions defined on $\mathbb{R}$. The measurements of Y are uncorrelated from the other ones, which is to say that the random variables, ε_i's, are uncorrelated, with unknown variance σ^2. The index i runs on all $1, .., n$. Due to the class of the functions f, which are defined linearly, and thus, through g parameters, the estimate of $f(x)$ is defined through the estimators of the coefficients of f on a basis, hence through g estimators. The choice of linear estimators with respect to the observations appears to be natural one, and enjoys optimality with respect to the variance through the Gauss Markov Theorem, taking into account that unbiasedness holds the model.

In this context, that the support of the design has cardinality, g is a necessary and sufficient condition to derive the Gauss Markov estimator of $f(x)$.

A design ξ is defined by

$$\xi := (x_j, n_j)_{j=0}^{g-1}$$

where the *nodes* x_j are the environment points of measurement, and n_j denotes the number of replications of the measurement Y in environment x_j.

We now introduce the words interpolation and extrapolation.

When x is in $(-1, 1)$ and when measurements are feasible in any neighborhood of x then x is an interpolation point. We will exclude the trivial case that measurements can be performed at x itself. On the contrary, when there exists a neighborhood V of x (which may belong to $[-1, 1]$ or not) such that no measurement can be performed on V, then x is an extrapolation point. Designs performing minimal MSE for the estimate of $f(x)$ within a class of given estimators are called optimal designs (interpolation designs when x is an interpolation point, extrapolation design in the second case; our definition may differ from others, f.i. Kiefer and Wolfowitz, and also Spruill sometimes use the expression "interpolation" when x in $(-1, 1)$ has a neighborhood where no observation can be performed, a case which we call extrapolation.

To present the guidelines and the flavor of this book, we start with a brief glance on how the theory of optimal designs meets the theory of the uniform approximation of functions. We, thus, start with some insight into Chapter 4 and the following chapters.

The interpolation problem amounts to building an optimal design, where the maximal variance of the estimate of $f(x)$, as x runs in $[-1, 1]$, is minimal. The optimal design thus solves this minimax problem pertaining to the variance of the estimator of the regression function.

The extrapolation problem amounts to building an optimal design for the estimation of $f(x)$, where x is an extrapolation point. This is a local problem; and in this case we do not consider designs with uniform optimality properties with respect to x. The criterion for optimality is the variance of an estimator of $f(x)$ within the class of linear unbiased estimators.

REMARK.– In the following, the interval where x lays may be chosen as $[-1, 1]$ for convenience. Changing from $[-1, 1]$ to any interval $[a, b]$ is easy.

I.2. A generic case: the Hoel and Levine extrapolation scheme and the uniform interpolation design of Guest

Chapter 4 introduces the basic ingredients in design planning. The function f, which is a polynomial with degree less than or equal to $g-1$, may be written on any basis of polynomials. The first ingredient is a choice of a basis which depends explicitly on the nodes of the design, say $x_0, .., x_{g-1}$, where the measurements will be performed, which, at this stage, are undefined. This basis is chosen as the one formed by the elementary Lagrange polynomials on $\mathbb{R}$, defined by

$$l_j(v) := \prod_{\substack{i \neq j \\ i=0}}^{g-1} \frac{v - x_j}{x_j - x_i}.$$

The Gauss Markov $\widehat{f(x)}$ estimator of $f(x)$ is a linear combination of the $\overline{Y_j}$s, the empirical mean of all measurements on the node x_j. The coefficient of $\overline{Y_j}$ is the elementary Lagrange polynomial l_j evaluated for $v = x$.

The interest of the choice of this basis lies in the fact that the variance of $\widehat{f(x)}$ takes a very simple form in terms of the $l_j(x)'$s. Indeed it turns out that

$$Var\widehat{f(x)} = \sigma^2 \sum_{j=0}^{g-1} l_j^2(x)/n_j, \qquad \text{[I.1]}$$

where $\sigma^2 := Var(\varepsilon)$.

The optimal design for the estimation of $f(x)$ solves the minimization of $Var\widehat{f(x)}$ over all choices of the g nodes, x_is in $[-1,1]$ and, above all, choices of the number of replicates of the measure $Y(x_i)$, say n_i, under a constraint on the total number n of such experiments.

The way in which the minimization of $Var\widehat{f(x)}$ should be performed can be imagined as a two-step procedure, using a simulation technique: simulate g nodes on $[-1,1]$; optimize the frequencies n_j/n given the nodes, using the Karush-Kuhn-Tucker theorem see [BAZ 06]. Then evaluate the variance of $\widehat{f(x)}$ through the above formula [I.1]. Iterate the simulation of the g nodes until you find a minimum value for $Var\widehat{f(x)}$. The important fact is that we have introduced the polynomial function $v \to Var\widehat{f(v)}$ for any design, and that the way to minimize the scalar $Var\widehat{f(x)}$ turns out to be a minimization problem for the value of a polynomial with known degree at point x when the polynomial takes a very special form, as given in [I.1].

The optimal design ξ should minimize [I.1] with respect to all nodes and all choices of the $n_j's$ under the constraint $n_0 + .. + n_{g-1} = n$. For a given set of nodes, the solution of the optimization step on the frequencies changes $Var\widehat{f(x)}$ into $q_{g-1}(x)$ given by

$$q_{g-1}(x) = \sum_{j=0}^{g-1} |l_j(x)|$$

when $-1 \leq x \leq 1$, and which, when $x < -1$ or $x > 1$, takes the form

$$q_{g-1}(x) = \sum_{j=0}^{g-1} (-1)^j \, l_j(x).$$

This quantity is thus the variance of $\widehat{f(x)}$ for fixed nodes $x_0, .., x_{g-1}$, optimized upon the number of replications.

Consider the mapping q_{g-1} defined outside $[-1, 1]$ by

$$v \rightarrow q_{g-1}(v) = \sum_{j=0}^{g-1} (-1)^j \, l_j(v), \qquad \text{[I.2]}$$

hence, a polynomial with degree up to $g - 1$, and consider $q_{g-1}(x)$ as the evaluation at point x of the variance of $\widehat{f(x)}$ obtained as the value of this polynomial function at point x.

We need to minimize [I.2] upon the nodes, hence to identify the polynomial $v \rightarrow q_{g-1}(v)$ which minimizes $q_{g-1}(x)$. Continuing q_{g-1} on $[-1, 1]$ it is readily seen that this polynomial should have value $(-1)^j$ when evaluated on x_j, and should have a derivative 0 on each of the $x_j's$; note further that, on $[-1, 1]$, q_{g-1} may assume negative values, and that $q_{g-1}(x)$ is the variance of $\widehat{f(x)}$ when x lays outside $[-1, 1]$.

The only polynomial which satisfies these requirements is the best uniform polynomial approximation with degree $g - 1$ of the null function on the interval $[-1, 1]$, as follows from the Borel–Chebyshev Theorem, hence independent of x. This proves that the support of the optimal design is obtained in a unique way independently of the location of x, if we decide to restrict its definition through the minimization of the sup norm of a polynomial when restricted to $[-1, 1]$. Note that, if we look for nodes through any other class of functions, we may lose uniformity of the support of the extrapolation design with respect to x, and there is no alternative general theory to define those designs.

We have built a bridge between the statistical theory of optimal design and the theory of the uniform approximation of functions. To complete the optimal design, a direct application of the theorem by Karush, Kuhn and Tucker yields

$$n_j = \left[n \frac{|l_j(x)|}{\sum_{j=0}^{g-1} |l_j(x)|} \right]$$

where [.] denotes the integer part function. Remarkably, the entire design does not depend on the function f as such, but only upon its degree as a polynomial. The design is robust within all functions f, which are polynomial functions with the same degree. The resulting design is the Hoel and Levine [HOE 64] design, and is central to Chapter 4.

Also, uniformity of the variance of the estimate of $f(x)$ can be achieved, for example when x is an interpolation point; in this case we want to obtain a design which minimizes the variance of $\widehat{f(x)}$ uniformly on $[-1, 1]$. This design is due to Guest [GUE 58], and its nodes are the Chebyshev points in $[-1, 1]$. This turns out to be a minimax problem with respect to the variance in $[-1, 1]$ where the maximum is performed on x and the minimum is performed on the design. Maximizing on x and minimizing on the frequencies of sampling for fixed nodes implicates that the optimal frequencies $\sigma^2 g/n$ are uniform, hence independent of the nodes. Direct optimization provides the optimal design. We also consider the relationship between the extrapolation design due to Hoel and Levine, which is of a local nature, and the interpolation design of Guest; assume that our aim is to obtain a uniformly optimal design on the entire interval $[c, 1]$ where $c < -1$. Since the variance of $\widehat{f(c)}$ is a decreasing function of c as $c < -1$, it readily follows the optimal design of Hoel–Levine at point c becomes the optimal Guest design on $[c, 1]$, which therefore has the Chebyshev points on $[-1, 1]$ as its nodes, for c far away from -1. A accurate discussion on the value of the variable c can be found in Levine's paper [LEV 66].

We have thought it useful to close the chapter devoted to these topics with an application in accelerated runs. The basic problem is written: how should experiments be performed, when the measurements are lifetimes of some item, and when the quantity of interest is the expectation of this lifetime? Examples are in classical context, including the Weibull and the logistic cases. Standard reparametrization leads to a regression model of the kind related above. A multivariate case is presented.

I.3. Extrapolation design in non standard cases, algorithms

Chapter 5 is also in the same vein, but opens up to explicit algorithms for the definition of the support of the optimal design; the definition of the problem makes it clear that those points are no longer the Chebyshev points.

The problem at hand is the extrapolation of the function f, a polynomial with known degree, $g - 1$, at point x which lies inside $[-1, 1]$ in the case when measurements can be performed only on $[-1, C] \cup [D, 1]$ where $C < x < D$. Henceforth, there exists a neighborhood of x, where no measurements are obtainable; x is an extrapolation point. The problem of finding the optimal design reduces to a version of Hoel-Levine type: choose the number of points of support of the design in $[-1, C]$ and in $[D, 1]$, in such a way to make the variance of $\widehat{f(x)}$ be the evaluation at point x of a polynomial with degree $g - 1$. Proceed as in Chapter 4 in order to optimize $Var\ \widehat{f(x)}$ in a two step procedure, defining the frequencies of sampling precisely as in Chapter 4. This yields the second step, where we optimize on the nodes. The properties of the resulting polynomial which describes $Var\ \widehat{f(x)}$ conditioned on the nodes make it similar to the Chebyshev polynomial. The resulting points of oscillation of this polynomial, which do not belong to (C, D), determine the design. A variant of the algorithm due to Remez, which is described at length in Chapter 2, provides the design; this treatment somehow follows the approach by Spruill [SPR 87a]. Hoel, in [HOE 65a], which came before Spruill's work, considered a one-dimensional case and the corresponding problem in two dimensions for some peculiar class of polynomials, and does not make use of the theory of uniform approximation, as done by Spruill. However, Spruill's approach is adequate for the generalization to the case when the observation set consists of a finite union of compact sets; this is developed in Chapter 5. We emphasize that also Hoel's solution for the optimal design is of a numerical nature; he also provides the following interesting example, proposed in [KAM 61]: for a given set of $n + 1$ distinct points, and for $n + 1$ real numbers $0 = y_0 < ... < y_i < ... < y_n = 1$, there exists a unique collection of points $(x_0, .., x_n)$ and a unique polynomial P with degree n such that $P(x_i) = y_i$ and $P'(x_i) = 0$. Kammerer proposes an iterative method for the determination of P.

For a generalization for a finite union of compact sets, we refer to the following works: [LEB 69, GER 88, PEH 99, PEH 01, PEH 03, PEH 09, PEH 11, TOT 11], a list which is not exhaustive.

Finally, Chapter 6 deals with the sensitivity of the design with respect to various types of uncertainty; the first one pertains to the measurement approximation with respect to the nodes. We assume that no uncertainty with respect to the values of the x_is holds. Uncertainty of the input pertains, therefore, on the measurements themselves, namely due to the random effect ε. Small input errors may propagate drastically to the Lagrange interpolation polynomials, and therefore, on the estimators. Since a large sample size reduces the effect of the input errors we consider the precision of the estimator with respect to the sample size using concentration type inequalities; this yields to a discussion on the assumptions to be made on the error factors in the model; both cases when these errors have bounded or unbounded support are considered. Bounds on n are obtained, which ensure a given upper bound on the resulting error of estimation. In this section we take the interested reader to a path through some real analytic models, in order to motivate the infinite dimensional setting, considered here through finite dimension approximation due to classical methods in real analysis.

In the same Chapter 6, we propose a brief account of the relative error of the Lagrange interpolation polynomial, using a generalized Borel–Chebyshev Theorem; given the relative errors on the $f(x_i)$'s we propose a calculation of the resulting relative error on the Lagrange interpolation. This result may be used, for example, for simulation experiments to determine the minimal sample size which generate a given accuracy of the estimator.

We present three appendices, each on technical nature in analysis; they contain material which either adds to the proofs in the main core of the book, or presents additional notions which we found both useful and interesting. The material included in those appendices exceeds the useful tools of the present volume; they are, however, necessary when extending the present setting to more general models than polynomials.

I.4. Uniform approximation of functions, an outlook

A summary of the interaction between optimal designs and the theory of the approximation of functions can now be stated.

Under the hypothesis that f is a polynomial with known degree, the design is naturally suited on the variance criterion, which is a polynomial with known degree; the basic situation exemplified in Chapter 4 shows that the design relies on nodes which are the roots of this polynomial; in turn, changing the nodes changes the variance, so that all this turns to the problem of the approximation of the null function by a polynomial with known degree.

It appears that the approximating function should, at times, satisfy a number of constraints, typically that it should take prescribed values at some points, called the nodes. The natural setting turns out to be the interpolation scheme, a concept which stems from the Lagrange interpolation procedure, and which amounts to considering a basis of polynomials with prescribed values at the nodes. Chapter 3 is devoted to this topic introducing the Chebyshev nodes, which, furthermore, are the optimal measurement points in the Hoel–Levine design. The role of the nodes in the error of approximation is considered with care, and yields to a discussion of the Runge phenomenon, which sheds light on the role of the Chebyshev nodes for interpolation problems in cases when f is not a polynomial function, or is a polynomial with unknown degree. Approximate f by a Lagrange interpolating polynomial with a given degree. The substitution of f by this function induces a bias on the estimator of

$f(x)$ at any x, in direct relation with the error due to the approximation. We state in Chapter 3 that this error can be reduced through the tuning of the interpolation nodes, for smooth functions f. Also, the optimal nodes pertaining to the approximation step are precisely the Chebyshev nodes; this result follows a close analysis of the role of the nodes in the convergence of the Lagrange interpolation scheme; they also define the support of the Hoel–Levine optimal design. In the same realm of questions, we consider some issues in relation to the numerical instability of the Lagrange scheme, through the results of Erdös [ERD 61] and of Bernstein [BER 18]. In the same chapter, we will discuss the relative advantages of the spline approximations and the Lagrange ones.

We close this Introduction with a brief (non-exhaustive) bibliography, which covers both the material in the present volume, and some more advanced topics to be discussed in a companion treatise.

I.5. A general bibliography

The first important contribution to the theory of optimal designs is due to Kristine Smith [SMI 18], who introduced a concept, to be called later on by Kiefer and Wolfowitz a G-optimal design [KIE 58] (we owe to Kiefer, the alphabetic denomination of the designs). In her paper, Smith evaluates the G-optimal designs for the polynomial regression in one variable and low degree (less than 6), with independent, equally distributed, homoscedastic and uncorrelated errors. Later on, various designs have been defined through more general criteria for optimality, according to the various problems at hand. Wald in 1943 [WAL 43] suggested the comparison of designs using the values of non-centrality parameters for tests of hypotheses about all parameters. This criterion is impractical, since it depends on the values of the unknown parameters. As a practical replacement he was led to the generalized variance, equivalent to comparing the determinants of the information matrix, and so introduced D-optimality. The term D-optimality used to designate Wald's criterion was introduced by Kiefer and Wolfowitz in the 1950s. In 1952, Elfving [ELF 52] obtained a fundamental result, known as Elfving Theorem, and obtained an optimal design, for the so-called A-criterion. In 1970 [WYN 70] and 1972 [WYN 72], Wynn proposed a number of results pertaining to D-optimal designs. In 1976, Studden and Tsay [STU 76] provided a procedure to calculate the optimal design through the algorithm of Remez.

I.5.1. *Interpolation and extrapolation designs, analytic and algorithmic solutions*

We now consider the chronology in relation to the interpolation and the extrapolation designs.

Guest [GUE 58] determined the G-optimal design for the polynomial model with one variable with arbitrary known degree. He, thus, generalized Smith's result under the same hypotheses on the error.

In 1959 [KIE 59], Kiefer and Wolfowitz characterized a wide class of designs whose support is the Chebyshev set of points. Hoel and Levine in 1964 [HOE 64], determined the optimal extrapolation design for the polynomial regression. The support of this design is the set of Chebyshev points.

In 1965, Hoel determined the minimax design for the polynomial case in two dimensions [HOE 65a]. Also in 1965, Kiefer and Wolfowitz generalized the result of Hoel and Levine to the Chebyshev regression model [KIE 65a]. In 1966, Hoel provided a simplified version of the results obtained by Kiefer–Wolfowitz for the Chebyshev regression [HOE 66]. Also in 1966, Levine [LEV 66] discussed the relationship between the extrapolation design by Hoel and Levine and the design with uniform minimal variance.

Kiefer and Wolfowitz [KIE 64] and [KIE 65b], consider regression problems where observations may be taken at points in a set X which does not coincide with the set Z on which the regression function is of interest. They consider the problem of finding a design (allocation of observations) which minimizes the maximum over Z of the variance function (of estimated regression). Specific examples are calculated for one-dimensional polynomial regression when Z is much smaller than or much larger than X. A related problem of optimum estimation of two regression coefficients is studied.

In 1965, Hoel [HOE 65a] stated that a solution is given to the problem of where to choose $k + 1$ points and what weights to assign to them to obtain the minimum variance of what may be called an interior extrapolated value of a polynomial of degree k. It is assumed that observations can be taken in the interval $[-1, 1]$, except for a subinterval (α, β) located in its interior and within which the extrapolation occurs. This one-dimensional solution is then used to solve the corresponding two-dimensional problem, for a certain class of polynomials, in which it is desired to extrapolate inside a rectangular region that lies inside a larger rectangular region and within which observations can be taken, exclusive of the interior region. In addition, the two-dimensional exterior extrapolation problem is solved for the same class of polynomials as those used for interior extrapolation. In the same year, Hoel [HOE 65a] published a generalization of the Guest design in two dimensions.

In 1968, Studden [STU 68] considered the work by Kiefer and Wolfowitz [KIE 59], where they characterized the optimal design supported by the Chebyshev set of points. In his paper, it is shown that the optimal design for estimating any specific parameter is supported by one of two sets of points for Chebyshev systems with certain symmetry properties. Different proofs of the Kiefer–Wolfowitz results are also presented. These results were extended by Kiefer and Wolfowitz [KIE 65b] to cover non-polynomial regression problems involving Chebyshev systems and the large class of designs supported by the Chebyshev points was characterized.

In Herzberg and Cox [HER 72], designs were proposed, whose role is to minimize the effect of random error in interpolation or extrapolation. The case most intensively studied is the one for first-degree polynomial models in k factors, although some results are also given for second-degree models. Emphasis is put on designs giving minimum variance at a particular point and satisfying additional requirements leading to good estimates of parameters.

In 1979, Galil and Kiefer published a result in relation to extrapolation designs; see [GAL 79].

For the extrapolation problem with $k+1$ regression functions $\varphi_0, \cdots, \varphi_k$, Hoel proposed a sufficient condition, in terms of $\varphi_0, \cdots, \varphi_k$ for every optimal design to be supported by at most $k+1$ points. This complements the conditions on the existence of optimal designs with $k+1$ supporting points. Hoel also studied regression systems for which optimal extrapolation designs require exactly $k+1$ points [HOE 81].

Dette, Heiligers and Studden in [DET 95] considered, in the context of the linear regression, the geometric structure of a class of minimax optimality criteria containing Elfving's minimax and Kiefer's Φ_p-criteria as special cases. They showed that the optimal designs with respect to these criteria are also optimal for $A'\theta$, where A is any vector of a generalized Elfving set. The results explain the particular role of the A- and E-optimality criterion; and are applied for determining the optimal design with respect to Elfving's minimax criterion in polynomial regression up to degree 9.

We now come to the various contributions of Spruill.

In 1984, Spruill [SPR 84a] studied an optimal design for extrapolation which is associated with a biased estimator (the minimax estimator of Speckman). Some concomitant questions of optimal design are investigated. In 1985 [SPR 85], Spruill explored the robustness of the Hoel–Levine design. In extrapolating a function which is close to being a polynomial, the least squares estimator combined with the Hoel–Levine optimal design is shown to perform well, in terms of mean square error, when compared with an optimal spline extrapolator; in [SPR 87b] Spruill provided an optimal extrapolation design for the derivatives of the regression function. In 1987, Spruill [SPR 87a] considered the problem handled by Hoel [HOE 65b] in quite a different way and characterized optimal designs for interpolation both for the unbiased estimation of a polynomial mean and for the minimax linear estimation of more general means. In this work, observations are assumed to be available only on disjoint intervals $[A, B]$, $[D, E]$, and optimal designs are sought for the minimization of the variance at a single point c in $[B, D]$. Spruill, in 1990 [SPR 90a], provided an optimal interpolation design for the estimation of the value of a linear functional of an unknown polynomial mean. For example, optimal designs are given for the estimation of the average value of a multivariate polynomial over a sphere of radius $a < r$, when observations are available only in the region $r \leq \parallel x \parallel \leq R$. Finally in 1990 [SPR 90b], Spruill considerd the choice of the degree of a polynomial in a regression problem, following Anderson [AND 62], who proposed a test procedure which is optimal in some sense, for a given design. Optimization upon the design in Anderson's result for the criterion of the power of the test is the contribution of Spruill. This yields to an experimental design which places roughly equal masses close to the oscillation points of the appropriate Chebyshev polynomial.

The optimal design for the polynomial regression with unknown degree was also considered by Dette, Holger and Studden [DET 95]. This paper considers the problem of determining efficient designs for polynomial regression models when only an upper bound for the degree of the polynomial is known by the experimenter before the experiments are carried out. The optimality criterion maximizes a weighted mean of the relative D-efficiencies in the different models. The optimal (model robust) design is determined completely, in terms of its canonical moments which form the unique solution of a system of nonlinear equations. The efficiency of the optimal designs with respect to different criteria is investigated by several examples.

From the 1980s, robustness properties of extrapolation designs have been of interest. We may cite in this respect the following works: [HUA 88, DET 96], [FAN 99a, FAN 99b, XU 06, WIE 08, XU 09] and [XU 14].

Extrapolation designs under misspecification or approximation schemes are considered in [FAN 00] and [BRO 07]. Fang adopted a minimax approach, which determines an optimal design to minimize the maximum value of the integrated mean-square prediction error (IMSPE), with the maximum being evaluated over the departures from the model. It turns out that, in straight-line and quadratic regression, the minimax extrapolation designs are the same as the minimum integrated variance (MIV) extrapolation designs, which assume that the models are correct. These designs coincide with the $D1$-optimal designs if the extrapolation interval becomes sufficiently large. For cubic polynomial regression, some comparisons of the performance of the proposed designs with uniform and MIV designs are given. Broniatowski and Celant [BRO 07] proposed an extrapolation design when the regression function is a real analytic function. A control of the bias is discussed. This bias is due to the substitution of the regression function by a polynomial with given degree.

In another paper, Fang [FAN 03b] considered a case with a contaminated model. He considers minimax extrapolation designs for a multivariate linear model where the contamination term with respect to the linear part is only known through its L_2 norm.

He decomposes the mean integrated predictive error of the adjusted value in a bias-variance form. Given a bound on the bias, he obtained the design which minimizes the corresponding variance.

In 2012, Broniatowski, Celant, Leoni and di Battista [BRO 12] considered the control of error terms for some extrapolation designs.

I.5.2. *Elfving's contribution*

The characterization by Elfving provides explicit designs for linear forms of the parameters, called c-optimal designs, developed in one or two dimensions in his paper. Elfving's approach makes use of some concepts in convex geometry (leading to the so-called Elfving set); this geometric approach to the characterization of the optimal designs has been extensively studied in a number of papers, see among them [ELF 54]. Elfving's work has been continued by Chernoff [CHE 53]. In 1954, de la Garza [GAR 54] and [KIE 59] proved that optimal designs only require us to consider information matrices of designs with trials at $k + 1$ points where k is the number of parameters in the regression, these including the extreme points of the observation set. Ehrenfeld [EHR 55] introduced a new criterion, which consists of the maximization of the minimal eigenvalue of the information matrix.

Elfving's theorem is the starting point of a broad class of works, aiming at its application in a number of applied situations. His aim is to estimate a linear form of the parameters of the model, with known coefficients, defined by a vector c. These linear forms are called c-forms. Their extensive study is the core of the companion volume of the present one.

Studden in [STU 71] provided a generic matrix formulation of the Elfving theorem, substituting the criterion of the minimal variance by a generic quadratic risk. Dette [DET 93] extend the Elfving theorem to the D-criterion.

The extension of the Elfving theorem to the heteroscedastic regression is due to Dette and Holland-Letz [DET 09]. These authors considered the case when the nonlinear regression includes a variance and a mean which are parametric functions of the explanatory variables. The c-optimal design problem, namely the optimal design pertaining to the estimation of the c-form, is investigated in the case when the parameters of both the mean and the variance function are of interest. A geometric characterization of c-optimal designs in this context is presented, which generalizes the classical result of Elfving for c-optimal designs. As in Elfving's famous characterization, c-optimal designs can be described as representations of boundary points of a convex set.

Fan and Chaloner in 2003 [FAN 03a], and Bartroff in 2012 [BAR 12], in the geometric standpoint of Elfving, using the equivalence theorem (see below), extended some c-optimal designs to some specific contexts.

In addition, the following papers illustrate the role of the Elfving theorem: Chernoff in 1999 [CHE 99] analyzes its role in the theory of the optimal designs. Fellman in 1999 [FEL 99] also provides an overview of Elfving's contribution in this field. An exhaustive review of Elfving's contribution can be found in the paper by Nordström [NOR 99].

Some other basic results in the theory of optimal designs are the equivalence theorem.

In 1958, Hoel [HOE 58] showed that in some cases the D-criterion introduced by Wald and the G-criterion introduced by Smith provide the same solution. Two years later, Kiefer and Wolfowitz [KIE 60] proved that this equivalence holds quite generally. This fact is known as the equivalence theorem. Indeed, it is the tool which has been used by Kiefer and Wolfowitz to prove the D-optimality of some designs.

Box and Lucas in 1959 [BOX 59] used the Chernoff, Elfving and Wald approaches to find locally D-optimum designs for the single response of nonlinear models arising in chemical kinetics and give a geometric interpretation of D-optimality which follows from that of Elfving. The designs of Box and Lucas are locally optimum and generalize the notion of D-optimality to nonlinear designs.

Kiefer in [KIE 74] generalized the equivalence theorem. He considered a general optimality criterion Φ; criteria equivalent to Φ-optimality are obtained under various conditions on Φ. Such equivalent criteria are useful for analytics or machine computation of Φ-optimum designs.

The contribution of Kiefer and Wolfowitz in the theory of optimal designs is quite important. We quote that Kiefer proposed to consider the experiment as a probability measure with which a matrix (the information matrix, which is in connection with the matrix of moments) is associated. The class of those matrices is a convex set, and its properties yield the equivalence theorem.

From the equivalence theorem follows a study of the algorithms which lead to the calculation of the D-optimal designs; see Fedorov and Malyutov [FED 72], Wynn [WYN 70], Mitchell [MIT 74], Silvey and Titterington [TIT 73]. Whittle in 1973 [WHI 73] generalized the equivalence theorem to the nonlinear case. He proved that the various cases of optimality considered so far, as C-optimality, D-optimality and D_s-optimality can be formulated in terms of maximizing a concave criterion function Φ of the design. He obtained a new and different characterization of D-optimality. His formulation also has implications for iterative algorithms of the type suggested by Wynn. Wynn in 1975 [WYN 75] provided conditions for the convergence of some algorithm which provides the optimal design. Finally Wu and Wynn [WU 78] also considered such problems. We also may cite the paper by Fidalgo and Rodrìguez-Dìaz in 2004 [FID 04], where a computational procedure for finding c-optimal designs using the Elfving method for more than two dimensions is provided. It is suitable for any model, achieving an efficient performance for three or four dimensional models.

Ulterior works to the classical theory (referring to Kiefer and Wolfowitz approach) include Hoel's work [HOE 62]. In this paper Hoel showed that the choice of x values that minimize the generalized variance of the estimates of the coefficients in polynomial regression also minimize the maximum of the generalized variance of the estimates of an arbitrary number of ordinates on the regression curve. This optimum spacing is also optimal with respect to the estimation of the coefficients of the first derivative of the polynomial regression function.

Karlin and Studden [KAR 66] discussed a number of results concerning the geometric theory of the optimal design of experiments. They consider the estimation of the $\mathbf{c}$-linear form $< \mathbf{c}, \theta >$, where θ is the vector of unknown regression coefficients. They provided a proof of the Elfving theorem in this case and characterized the optimal design for estimating $< c, \theta >$. They showed that the Hoel and Levine's result on extrapolation, and some of its extensions, can be deduced from Elfving's result. They also considered a generalization of the Guest design and determined several explicit minimax designs for certain cases where the regression functions are related to classical orthogonal polynomials.

I.5.3. *Some books in relation to the present one*

We now close this bibliography with a survey on books pertaining to optimal designs.

The first book which provides an overview in the area is by Fedorov and Malyutov [FED 72]. In 1980, Silvey [SIL 80] published a book which presented the equivalent theorem in a very accurate and clear way. We next find the book by Pázmann [PÁZ 86].

Atkinson and Donev [ATK 07] produced a book with applications to engineering. In 1993, Pukelsheim's book [PUK 93] shed light on the relationship between the optimal designs and convex programming. The second edition of Pukelsheim's book, a reference text, was issued in 2006.

The book by Schwabe [SCH 96] described optimal designs in multifactorial models. This monograph is concerned with a wide variety of problems on optimal experimental designs for multivariate regression models. Written clearly and with a high precision, it contained a large number of mathematically deep results in the area of (non-)optimality of the so-called product experiment designs in estimating parameters in multifactor models. Müller [MÜL 97] in 1997 published a book which linked the two distinct areas of robust statistics and experimental design.

In 2005, the book by Berger [BER 05] was the first to catalog the application of optimal design to real problems, in computer-aided methods in optimal design and operations. Papers from the Bilateral Workshop of British and Lithuanian Scientists [BOG 06] cover different topics on optimal design and operations with particular emphasis on chemical engineering applications.

Also in 2006, Melas [MEL 06] introduced the basic concepts needed in optimal experimental design theory. A general theory of the functional approach to studying locally D-optimal and maximin efficient designs was also given in the same. The theory was then applied to various functional aspects of the regression function. Locally E-optimal designs for a class of nonlinear models were considered too.

In 2009, Stidham and Shaler [STI 09] produced a book at the frontier between optimal designs and operation research. The invention of dynamic programming by Richard Bellman provided a coherent framework for studying optimal control in queueing systems. Their book concentrated on the optimal design, i.e. setting parameters of the system before putting it into operation, and left the dynamic control to a subsequent volume.

In 2013, Pronzato and Pázman [PRO 13] published a book which provided a mathematically rigorous introduction to designs for nonlinear models.

In 2014, Çela *et al.*'s [CEL 14] book focused on the design of special control and scheduling algorithms based on system structural properties as well as on the analysis of the influence of induced time-delay on system performances. It discussed the optimal design of distributed and embedded control systems with respect to communication and calculation–resource constraints, quantization aspects, and potential time delays induced by the associated communication and calculation model. This book is of great interest for researchers in signal processing and in optimal control.

PART 1

Elements from Approximation Theory

1

Uniform Approximation

1.1. Canonical polynomials and uniform approximation

For fixed n, let $\mathcal{P}_n(I)$ denote the class of all polynomials with a degree equal to or less than n defined on I, a compact subset of $\mathbb{R}$, and let $P(x) := a_0 + ... + a_n x^n$. When set I is clearly identified by the context, we will write $\mathcal{P}_n$ instead. Some coordinates of the vector of coefficients $(a_0, ..., a_n)$ can take value 0.

For a continuous function f defined on I denote:

$$R := f - P$$

which is a continuous function on I, as is $|R|$.

Applying the Weierstrass theorem, it follows that $|R|$ attains its maximal value in I, for at least one point x. We denote:

$$\mu = \mu(a_0, a_1, ..., a_n) := \max_{x \in I} |R(x)| \geq 0$$

which yields to define

$$E^* := \inf_{(a_0, a_1, ..., a_n)} \mu(a_0, a_1, ..., a_n), \qquad [1.1]$$

the minimum value of the uniform error committed substituting f by a polynomial in $\mathcal{P}_n$. This lower bound exists.

DEFINITION 1.1.– A polynomial P^* in $\mathcal{P}_n$ with coefficients $(a_0^*, ..., a_n^*)$ such that

$$\|f - P^*\|_\infty := \sup_{x \in I} |f(x) - P^*(x)| = E^*,$$

where E^* is defined in [1.1] and is called a best approximating polynomial of f with degree n in the uniform sense.

This best approximation thus satisfies

$$P^* := \arg \inf_{P \in \mathcal{P}_n} \sup_{x \in I} |R(x)|. \qquad [1.2]$$

The polynomial P^* may be of a degree less than n.

Define

$$R^*(x) := f(x) - P^*(x) \qquad [1.3]$$

and

$$E^* = \max_{x \in I} |R^*(x)|. \qquad [1.4]$$

REMARK 1.1.– The Weierstrass theorem for continuous functions has been proved in many different ways, stemming from the original proofs by Weierstrass and Runge, followed by Bernstein. All proofs identify a specific sequence of polynomials; for example the family of Bernstein polynomials, the Lagrange interpolation polynomials or polynomials defined by convolutions. The difference between these various approaches lies in the rate of convergence which clearly differs from one sequence to another, yielding different approximation rates for the same degree. The quantity

$$E^* := E_n^*(f) := \inf_{P_n \in \mathcal{P}_n} \sup_{x \in [a,b]} |f(x) - P_n(x)|$$

which measures the adequacy of the approximation of the continuous function f is called the convergence rate.

1.2. Existence of the best approximation

This section answers the question of attainment, for the least uniform error, when approximating f by a polynomial in $\mathcal{P}_n$.

THEOREM 1.1.– Let f be a continuous function defined on I. For any integer n there exists P^* in $\mathcal{P}_n$ such that $\|f - P^*\|_\infty = \inf_{P \in \mathcal{P}_n} \|f - P\|_\infty$.

PROOF.– Since $\mathcal{P}_n$ is a linear space, the null polynomial $\mathbf{0}$ belongs to $\mathcal{P}_n$. Therefore if P^* exists

$$\|f - P^*\|_\infty \leq \|f - \mathbf{0}\|_\infty = \|f\|_\infty$$

holds.

Henceforth, P^* should belong to the set

$$\overline{S(f, \|f\|_\infty)} \cap \mathcal{P}_n := \left\{ g \in \mathcal{C}^{(0)}(I) : \|g - f\|_\infty \leq \|f\|_\infty \right\} \cap \mathcal{P}_n.$$

This is a closed and bounded subset in $\mathcal{P}_n$. Since the dimension of $\mathcal{P}_n$ is finite, $\overline{S(f, \|f\|_\infty)} \cap \mathcal{P}_n$ is a compact set.

It is therefore enough to prove that the mapping

$$\Phi : (\mathcal{P}_n, \|.\|_\infty) \to (\mathbb{R}, |.|), P \mapsto \Phi(P) := \|f - P\|_\infty$$

is continuous.

From proposition A1.1 in Appendix 1, for Q in $\mathcal{P}_n$,

$$|\Phi(P) - \Phi(Q)| \leq \|P - Q\|_\infty,$$

proving the claim.

The function Φ, thus, is continuous; it has at least a minimal value P^* in the compact set $\overline{S(f, \|f\|_\infty)} \cap \mathcal{P}_n$. □

1.3. Characterization and uniqueness of the best approximation

The proof of the uniqueness theorem heavily relies on a theorem by Chebyshev which explores the number of changes of the sign of the error, which we state now. Consider P^* and E^* as defined in [1.2] and [1.4]. The set I is now an interval

THEOREM 1.2 (Borel–Chebyshev).– A polynomial P^* in $\mathcal{P}_n$, is the best uniform approximation of a function f in $\mathcal{C}^{(0)}([a, b])$, if and only if the function $x \to R^*(x)$ equals E^* with alternating signs for at least $n + 2$ values of x in $[a, b]$.

The proof of this important result is postponed to the end of the present section.

We introduce the following definitions:

DEFINITION 1.2.– The $n + 2$ points defined in theorem 1.2 are called equioscilation points.

This class of points is a subset of the class of alternating points, defined by:

DEFINITION 1.3.– Given a function g, defined on some set X, a family of points $x_1, .., x_m$ in X is an alternating family of points for g if $|g(x_1)| = .. = |g(x_m)|$ and $sign(g(x_i)) = sign(g(x_{i+1})), 1 \leq i \leq m - 1$.

Theorem 1.2 asserts that the approximation of a function f in $\left(\mathcal{C}^{(0)}\left([a,b]\right), \|.\|_{\infty}\right)$ requires handling the parameters $A = \{a_0, ..., a_n\}$ which are the coefficients of P for a generic P in V. Reducing the distance $|f(x) - P(x)|$ on a point x in $[a,b]$ may result in an increase of the distance on another point. The optimal approximation holds when the maximal distance $\sup_{x\in[a,b]} |f(x) - P(x)|$ holds equal on $n+2$ points. Furthermore, this maximal distance should be attained with alternating signs.

Theorem 1.2 can be extended when f belongs to a normed space $(\mathbf{V}, \|.\|)$, instead of $\left(\mathcal{C}^{(0)}\left([a,b]\right), \|.\|_{\infty}\right)$. A special case is when,

$$(\mathbf{V}, \|.\|) = \left(\mathcal{C}^{(1)}\left([a,b]\right), \|.\|_{[a,b]}^{1}\right)$$

where $\|f\|_{[a,b]}^{1} := \|f\|_{\infty} + \|f'\|_{\infty}$ (with $x \in [a,b]$).

We now introduce the following definition.

DEFINITION 1.4.– The critical points of the function f are defined as the set:

$$\mathbf{Cr}\,(f) := \{x \in [a,b] : |f(x)| = \|f(x)\|_{\infty}\},$$

the set of points where, f reaches its maximal absolute values in $[a,b]$.

We now state the important uniqueness result.

THEOREM 1.3.– Let f be a continuous function defined on $[a,b]$. The best uniform approximating polynomial is unique.

PROOF.– In order to prove uniqueness, assume that two such polynomials of best approximation, say P and Q exist. Then

$$S := \frac{P+Q}{2}$$

belongs to $\mathcal{P}_n$ and for any x in $[a,b]$

$$\begin{aligned}|f(x) - S(x)| &\leq \frac{1}{2}|f(x) - P(x)| + \frac{1}{2}|f(x) - Q(x)| \\ &\leq \inf_{P\in\mathcal{P}_n} \|f - P\|_{\infty} =: E^{*}.\end{aligned}$$

Furthermore, the polynomial S is the best uniform approximation of f.

Consider a point z in $[a, b]$ where, by continuity

$$\left| f(z) - \frac{P(z) + Q(z)}{2} \right| = E^*.$$

It holds

$$E^* \leq \frac{1}{2} |f(z) - P(z)| + \frac{1}{2} |f(z) - Q(z)| \leq E^*$$

hence

$$\frac{1}{2} |f(z) - P(z)| + \frac{1}{2} |f(z) - Q(z)| = E^*.$$

Denote $\alpha := \frac{1}{2}(f(z) - P(z))$ and $\beta := \frac{1}{2}(f(z) - Q(z))$. It then holds

$$|\alpha + \beta| = |\alpha| + |\beta|.$$

Now $\frac{1}{2}|f(z) - P(z)| \leq E^*/2$ and so is $\frac{1}{2}|f(z) - Q(z)|$. Therefore, both equal $E^*/2$, i.e. $\alpha = \beta$. Therefore, $f(z) - P(z) = f(z) - Q(z)$ hence $P(z) = Q(z)$, which proves that z is the root of $P - Q$, a polynomial with degree less or equal n. Recall that S is a best uniform approximation of f. From Chebyshev–Borel's theorem, S admits $n+2$ maxima; therefore, we may find $n+2$ distinct points in $[a, b]$ similar to z. Since P and Q have degree less or equal n, they coincide. This closes the proof. □

1.3.1. *Proof of the Borel–Chebyshev theorem*

The proof obtained is through contradiction, assuming that the function $x \to R^*(x)$ takes h times the value E^* with alternating signs, with $h \leq n+1$. Under this assumption, we will construct a polynomial $Q^*(x)$ for which

$$\max_{x \in [a,b]} |f(x) - Q^*(x)| < E^*,$$

is a contradiction.

Let

$$x_1, ..., x_h,$$

denotes the h points where $|R^*|$ equals E^*.

Define

$$\varepsilon_i := sign\left(f\left(x_i\right) - P^*\left(x_i\right)\right) \in \{-1, 1\}.$$

For any i between 1 and h, it holds

$$f\left(x_i\right) - P^*\left(x_i\right) = \varepsilon_i E^*.$$

Consider the system of h linear equations with $n+1$ variables $a_0, ..., a_n$ defined through,

$$\begin{cases} a_0 + a_1 x_1 + ... + a_{n-1} x_1^{n-1} + a_n x_1^n = \varepsilon_1 E^* \\ a_0 + a_1 x_2 + ... + a_{n-1} x_2^{n-1} + a_n x_2^n = \varepsilon_2 E^* \\ \dots\dots\dots\dots\dots\dots\dots\dots\dots\dots \\ a_0 + a_1 x_h + ... + a_{n-1} x_h^{n-1} + a_n x_h^n = \varepsilon_h E^* \end{cases}.$$

Since, $h \leq n+1$, this system has at least one solution; if $h < n+1$, it has an infinite set of solutions.

Denoted by

$$a_0 = p_0, ..., \ a_n = p_n,$$

such a solution.

Since E^* is not zero, all the p_is cannot simultaneously take the value 0.

Using $(p_0, ..., p_n)$ define the following polynomial

$$\widetilde{P}(x) := \sum_{j=0}^{n} p_j x^j.$$

This polynomial shares a number of properties with $|R^*|$. First $\left|\widetilde{P}(x_i)\right| = |R^*(x_i)| = E^*$, for all $i = 1, 2, .., h+1$.

We now prove that a specific linear combination of $\widetilde{P}$ and P^* provides better uniform approximation of f with respect to P^*, which concludes the proof.

For any positive β, defined on $[a, b]$ the following continuous function

$$x \mapsto f\left(x\right) - P^*\left(x\right) - \beta\, \widetilde{P}(x).$$

The proof produces a value of β such that

$$\sup_{x\in[a,b]} \left| f(x) - P^*(x) - \beta\, \widetilde{P}(x) \right| < E^*. \qquad [1.5]$$

From a continuity argument, there exists some ϵ such that $0 < \epsilon < E^*$ with

$$|x' - x''| \leq \delta \Rightarrow \begin{cases} |R^*(x') - R^*(x'')| < \epsilon \\ \left| \widetilde{P}(x') - \widetilde{P}(x'') \right| < \epsilon \end{cases} \qquad [1.6]$$

when x' and x" both belong to $[a, b]$.

The proof aims to bound the supremum in [1.5], splitting $[a, b]$ as follows: we will consider neighborhoods $\mathcal{I}_\delta(x_i)$ of the points x_i where

$$\sup_{x\in[a,b]} |f(x) - P^*(x)|$$

attains its maximal value, and the complementary set of those in $[a, b]$.

STEP 1.– We bound $f(x) - P^*(x) = R^*(x)$ on neighborhoods of points x_i's, where the maximal error is attained with positive sign.

Consider any interval $\mathcal{I}_\delta(x_i) := \left(x_i - \frac{\delta}{2}, x_i + \frac{\delta}{2}\right)$ with length δ and center x_i, where

$$R^*(x_i) = E^*$$

meaning that x_i is a point where P^* reaches its maximal (positive) value.

Let $x' = x_i$ and $x'' = x \in \mathcal{I}_\delta(x_i)$. In the interval $\mathcal{I}_\delta(x_i)$, since

$$|R^*(x') - R^*(x'')| < \epsilon,$$

we deduce that for all x in $\mathcal{I}_\delta(x_i)$

$$R^*(x_i) - R^*(x) < \epsilon \qquad [1.7]$$

where we used the fact that for any x in $\mathcal{I}_\delta(x_i)$, $E^* \geq |R^*(x)|$.

Hence, from [1.7]

$$E^* - \epsilon < R^*(x).$$

Since $R^*(x) < E^*$, it also holds that

$$E^* - \epsilon < R^*(x) < E^*. \qquad [1.8]$$

STEP 2.– We bound $\widetilde{P}$ on neighborhoods of points x_i's, where the maximal error is attained with a positive sign.

Consider the polynomial $\widetilde{P}(x)$ on $\mathcal{I}_\delta(x_i)$, which satisfies

$$\widetilde{P}(x_i) = E^*.$$

By continuity

$$\left|\widetilde{P}(x') - \widetilde{P}(x'')\right| < \epsilon$$

for ϵ defined in [1.6].

This yields,

$$-\epsilon < \widetilde{P}(x_i) - \widetilde{P}(x) < \epsilon,$$

i.e.

$$E^* - \epsilon < \widetilde{P}(x) < E^* + \epsilon. \qquad [1.9]$$

STEP 3.– We bound $f(x) - P^*(x) - \beta\,\widetilde{P}(x)$ on neighborhoods of points x_i's, where the maximal error is attained with a positive sign.

For any positive ω, [1.9] is equivalent to

$$\omega(E^* - \epsilon) < \omega\widetilde{P}(x) < \omega(E^* + \epsilon). \qquad [1.10]$$

From [1.8] and [1.10]

$$E^* - \epsilon - \omega(E^* + \epsilon) < R^*(x) - \omega\widetilde{P}(x) < E^* - \omega(E^* - \epsilon).$$

Choose any positive ω, such that

$$\omega < \frac{E^* - \epsilon}{E^* + \epsilon}. \qquad [1.11]$$

With this choice, for any x in $\mathcal{I}_\delta(x_i)$,

$$0 < R^*(x) - \omega\widetilde{P}(x) < E^* - \omega(E^* - \epsilon).$$

STEP 4.– We do the same as in the three previous steps taken on the neighborhoods of points x_i's, where the maximal error is attained with negative sign.

Until now, we have considered the points x_i's in $[a,b]$ where $R^*(x_i) = E^*$. We use the same construction as above, using the points x_j for which $R^*(x_j) = -E^*$, defining similarly intervals $\mathcal{I}_\delta(x_j)$ with length δ and centered at x_j.

A similar argument provides

$$-E^* + \omega(E^* - \epsilon) < R^*(x) - \omega\widetilde{P}(x) < 0.$$

Henceforth, in any interval of the form $\mathcal{I}_\delta(x_i)$, where $|R^*(x_i)| = E^*$, it holds

$$\left|R^*(x) - \omega\widetilde{P}(x)\right| < E^* - \omega(E^* - \epsilon). \qquad [1.12]$$

This inequality is valid on $\cup_{i=1}^h \mathcal{I}_\delta(x_i)$.

STEP 5.– We bound $f(x) - P^*(x) - \omega\widetilde{P}(x)$ on the complementary set of $\cup_{i=1}^h \mathcal{I}_\delta(x_i)$ in $[a,b]$.

We now explore the behavior of R^* outside the collection of the intervals $\mathcal{I}_\delta(x_i)$ where it assumes its extreme values.

Let

$$A := [a,b] \setminus \left(\cup_{i=1}^h \mathcal{I}_\delta(x_i)\right).$$

On A, the mapping $x \to |R^*(x)|$ attains its maximal value m such that $m < E^*$.

Since $A \subseteq [a,b]$ it holds $m \leq E^*$. Furthermore $m \neq E^*$; indeed A and $\cup_{i=1}^h \mathcal{I}_\delta(x_i)$ are disjoint sets.

On A it holds,

$$|R^*(x)| \leq m. \qquad [1.13]$$

Let

$$M := \max_{x\in[a,b]} \left|\widetilde{P}(x)\right|.$$

We impose the additional condition on ω, defined in [1.10] through

$$0 < \omega < \frac{E^* - m}{2M}.$$

It then holds, for any x in A, that

$$\left|\omega \widetilde{P}(x)\right| < \frac{E^* - m}{2}. \tag{1.14}$$

Apply the triangle inequality to $\left|f(x) - P^*(x) - \omega \widetilde{P}(x)\right|$, to obtain

$$\left|f(x) - P^*(x) - \omega \widetilde{P}(x)\right| \leq |f(x) - P^*(x)| + \omega \left|\widetilde{P}(x)\right|.$$

From [1.13] and [1.14] it follows that on A

$$\left|R^*(x) - \omega \widetilde{P}(x)\right| \leq \frac{m + E^*}{2}.$$

STEP 6.– We assemble the previous results to obtain that $P^*(x) - \omega \widetilde{P}(x)$ improves on P^*, a contradiction.

From that shown above and [1.12] we obtain for any x in $[a, b]$

$$\left|R^*(x) - \omega \widetilde{P}(x)\right| < \min\left\{\frac{m + E^*}{2}, E^* - \omega(E^* - \epsilon)\right\}.$$

Furthermore, since $m < E^*$ and by the very choice of ω

$$\min\left\{\frac{m + E^*}{2}, E^* - \omega(E^* - \epsilon)\right\} < E^*,$$

we obtain, for any x in $[a, b]$.

$$\left|R^*(x) - \omega \widetilde{P}(x)\right| = \left|f(x) - P^*(x) - \omega \widetilde{P}(x)\right| < E^*.$$

This means that

$$Q^*(x) := P^*(x) + \omega \widetilde{P}(x),$$

approximates f with an error smaller than that obtained through P^*, a contradiction.

This closes the proof.

As done previously, through theorem 1.2, the best polynomial approximation P^* of f is characterized by the fact that the function

$$x \to f(x) - P^*(x)$$

takes its common extreme value on at least $n+2$ points, with alternating sign, hence equioscillating points.

1.3.2. *Example*

EXAMPLE 1.1.– Define the Chebyshev polynomial of first kind defined on $[-1, 1]$ by,

$$T_n(x) := \cos(n \arccos x).$$

This polynomial has as equioscillation points the set

$$x_i := \cos(i\pi/n), 0 \leq i \leq n,$$

see Appendix 2. The polynomial $x \mapsto x^n - \frac{2}{2^n} T_n(x)$ has degree n and equioscillates $n+1$ times on the above x_i's. Hence, it is the best uniform approximation of the null function on $[-1, 1]$. Note that this best approximating polynomial is defined by fixing its value to 1. Indeed, when approximating the null function, all approximations are defined up to a multiplicative constant. This, obviously, is not the case for the approximation of any other function.

REMARK 1.2.– More generally, let $f : [-1; 1] \to \mathbb{R}, x \mapsto f(x) := \sum_{j=0}^{n+1} a_j x^j$, with $a_{n+1} \neq 0$. We determine the polynomial $\mathrm{P}_n^* \in \mathcal{P}_n$, which approximates f in $[-1; 1]$. This results in P_n^* such that $(f - P_n^*)/a_{n+1}$ is the best polynomial approximation of the null function with a degree less or equal $n+1$ with value 1. Since this polynomial is $2^{-n} T_{n+1}$, it holds $f - P_n^* = 2^{-n} T_{n+1}$. Hence, $P_n^* = \sum_{j=0}^{n+1} a_j x^j - 2^{-n} a_{n+1} T_{n+1}$.

EXAMPLE 1.2.– Let $f(x) = 9x^4 - x$. The best uniform approximation with a degree less or equal to 4 on $[-1; 1]$ is $P_4^*(x) = 9x^4 - x - \frac{9}{8}\left(8x^4 - 8x^2 + 1\right) = 8x^2 - x - 1$.

EXAMPLE 1.3 (see [DZY 08]).– Let $f(x) = \sin x, x \in [-4\pi; 4\pi]$. Determine the polynomial $P_n^* \in \mathcal{P}_6$, corresponding to f in $[-4\pi; 4\pi]$. The function $\sin x$ alternates its sign on the points $x_j = -\frac{7}{2}\pi + k\pi$ and $k = 0, 1, ..., 7$. Let $\mathbf{0}_6 : [-4\pi; 4\pi] \to \mathbb{R}, x \mapsto \mathbf{0}_6(x) := 0$; then on $x_j \in [-4\pi; 4\pi], j = 0, 1, ..., 7$, it holds: $\sin x_j - \mathbf{0}_6 = (-1)^j \|\sin x_j\|_{\infty,[-4\pi;4\pi]} = (-1)^j$. Hence, form the Chebyshev–Borel theorem, the null polynomial $\mathbf{0}_6 \in \mathcal{P}_6([-4\pi; 4\pi])$ is the best uniform approximation of $\sin x$ in $\mathcal{P}_6([-4\pi; 4\pi])$.

EXAMPLE 1.4.– Determine in

$$\mathcal{P}_n(1) := \{P \in \mathcal{P}_n([a;b]), 0 < a < b : P(0) = 1\}$$

the best approximation of the null function. Intuitively, since $T_n\left(\frac{2x-a-b}{b-a}\right)$ is the best approximation of the null function on $[a;b]$, we may say that $T_n\left(\frac{2x-a-b}{b-a}\right)/T_n\left(\frac{-a-b}{b-a}\right)$ is the solution. Indeed, the constant $1/T_n\left(\frac{-a-b}{b-a}\right)$ makes $T_n\left(\frac{2x-a-b}{b-a}\right)/T_n\left(\frac{-a-b}{b-a}\right)$ evaluated at 0 equal to 1. We prove that $T_n\left(\frac{2x-a-b}{b-a}\right)/T_n\left(\frac{-a-b}{b-a}\right)$ is indeed the solution. By contradiction, if $Q \in \mathcal{P}_n$ and $Q(0) = 1$ is the solution then $T_n\left(\frac{2x-a-b}{b-a}\right)/T_n\left(\frac{-a-b}{b-a}\right) - Q_n(x)$ alternates values in at least $n+1$ points in $[a;b]$. This function thus has n roots in $[a;b]$ at least. On $x = 0$, $Q(0) = 1$. Hence $T_n\left(\frac{2x-a-b}{b-a}\right)/T_n\left(\frac{-a-b}{b-a}\right) - Q(0) = 0$. Hence $T_n\left(\frac{2x-a-b}{b-a}\right)/T_n\left(\frac{-a-b}{b-a}\right) - Q(x)$ has at least $n+1$ root. But this function has degree n, thus $T_n\left(\frac{2x-a-b}{b-a}\right)/T_n\left(\frac{-a-b}{b-a}\right) = Q(x)$ for all x.

REMARK 1.3.– In the example above, the minimal value of the functional $\|.\|_\infty$ is on the frontier of the unit sphere of $\mathcal{P}_n([a,b])$. Such points are called extreme points. See Chapter 5 for extensions.

EXAMPLE 1.5.– Let

$$f : [-1,1] \to \mathbb{R}, x \mapsto f(x) := x^4.$$

The best uniform approximation of f in $\mathcal{P}_3([-1,1])$ is given by,

$$P^*(x) := x^2 - \frac{1}{8}.$$

The derivative of the error function

$$R^*(x) := x^4 - x^2 + \frac{1}{8}$$

takes value 0 in $x_1 = 0, x_2 = \frac{\sqrt{2}}{2}$ and $x_2 = -\frac{\sqrt{2}}{2}$.

Furthermore

$$R^*(-1) = \frac{1}{8}, R^*(0) = \frac{1}{8}, R^*\left(\frac{\sqrt{2}}{2}\right) = -\frac{1}{8} \text{ and } R^*\left(-\frac{\sqrt{2}}{2}\right) = -\frac{1}{8}, R^*(1) = \frac{1}{8}.$$

Henceforth, there exist five points where the absolute value of the error is maximal. The error alternates sign.

2

Convergence Rates for the Uniform Approximation and Algorithms

2.1. Introduction

In this Chapter, the Borel–Chebyshev theorem is considered as well as the effective calculation of the minimax polynomial through the algorithm due to Remez. We will consider some examples too. We will also consider the rate of convergence of the uniform approximation, as well as the rate of convergence of the Remez algorithm. We will now state proof of the Borel–Chebychev theorem in relation to the uniform polynomial approximation to start begin.

2.2. The Borel–Chebyshev theorem and standard functions

A necessary condition for the Borel–Chebyshev theorem to hold is that the function f to be approximated should be continuous. We will consider a strengthening of this assumption, assuming a derivability assumption. We will see that this additional requirement provides a wealth of information on the approximating polynomial; namely, the frontier points, of the compact set where f is to be approximated, are equioscillation points for the error function, and differentiability of f provides the number of such equioscillation points.

Let us start with some definitions to start with.

DEFINITION 2.1.– Let $P^* \in \mathcal{P}_n$ be the minimax polynomial of the function $f : [a, b] \to \mathbb{R}$. The error function $R^* := f - P^*$ is standard if it has exactly $n + 2$ local extrema (minima or maxima), on $[a, b]$, where the frontier points a and b also belong to this set of points.

THEOREM 2.1.– Let $f \in \mathcal{C}^{(n+1)}([a,b])$ and $f^{(n+1)}(x) \neq 0$ for all $x \in [a,b]$. Then, R^* is a standard function.

PROOF.– From the Borel–Chebyshev theorem we know that R^* changes sign at least $n+2$ times in $[a,b]$. Hence by Rolle Theorem, $R^* := f - P^*$ has at least $n+1$ distinct roots in $[a,b]$. Denote $z_1, ..., z_{n+1}$ any subset of $n+1$ distinct such roots; it holds $f(z_i) = P^*(z_i)$ for $i = 1, ..., n+1$. Hence, P^* is defined as the Lagrange interpolation polynomial of f which goes through the nodes z_i's. It follows the theory of the Lagrange interpolation (to be developed in Chapter 3, section 3.2) that R^* coincides with the interpolation error function and therefore,

$$R^*(x) = \frac{f^{(n+1)}(\xi)}{(n+1)!} \prod_{i=1}^{n+1} (x - z_i)$$

with $\xi \in (a,b)$. By hypothesizing for all $x \in [a,b]$, $f^{(n+1)}(x) \neq 0$. With $x = \xi$ it holds that $f^{(n+1)}(\xi) \neq 0$. Hence, R^* has exactly $n+1$ roots; hence, $n+2$ points of equioscillation in $[a,b]$, including a and b; the degree of the derivated polynomial of R^* is, indeed, n. In results that R^* has at most n equioscillation points inside (a,b). Since it has at least $n+2$ such points on $[a,b]$, the frontier points $\{a,b\}$ belong to the set of equioscillation points of R^*. □

We explore some examples.

EXAMPLE 2.1.– Let f be a continuous function on $[a,b]$ with continuous second derivative on (a,b). Determine the best uniform approximation of f in $\mathcal{P}_1$. Denote $P(x) := \theta_1 x + \theta_2$ this solution. Hence, by the Borel–Chebyshev theorem there should be at least $\deg(P^*) + 2 = 3$ distinct points, say $t_1 \neq t_2 \neq t_3$, in $[a,b]$ where

$$\begin{cases} R^*(t_1) = \epsilon E^* \\ R^*(t_2) = -\epsilon E^* \\ R^*(t_3) = \epsilon E^* \end{cases} \qquad [2.1]$$

with $\epsilon \in \{-1, 1\}$, $E^* := \|R^*\|_\infty$ and $R^* := f - P^*$. We will provide t_1, t_2, t_3 and E^*, θ_1 and θ_2.The function $R^* := f - P^*$ has at least three stationary points in $[a,b]$. Assuming f strictly convex on $[a,b]$ (i.e. $f^{(2)}(x) > 0$ for all $x \in [a,b]$), it holds that

$$\begin{cases} (R^*)'(x) = f'(x) - \theta_1 \\ (R^*)''(x) = f^{(2)}(x) > 0 \end{cases}.$$

Hence,

$$(R^*)'(x) = 0 \text{ when } f'(x) = \theta_1.$$

Since, $f^{(2)}(x) > 0$ for all $x \in (a,b)$, the equation $f'(x) = \theta_1$ has at least one solution which is the abscissa of a minimum of R^*. Furthermore, since R^* is strictly convex,

$$f'(x) = \theta_1$$

has a unique solution $x^* \in (a,b)$, which is the unique absolute minimum of R^*. Denote

$$m := \min_{x\in[a,b]} R^*(x).$$

Assign $x^{**} \in (a,b)$, some other absolute minimum value. Then $[x^*, x^{**}] \subset [a,b]$.

By convexity for all ζ in $[0,1]$

$$R(\zeta x^* + (1-\zeta)x^{**}) < \zeta R(x^*) + (1-\zeta)R(x^{**}) = \zeta m + (1-\zeta)m = m.$$

We may thus define,

$$x^{***} := \zeta x^* + (1-\zeta)x^{**} \in (a,b)$$

with

$$R^*(x^{***}) < m$$

which cannot hold. Therefore, x^* is the only absolute minimizer of R^* in (a,b).

Since P is the minimax polynomial, it follows from the Borel–Chebyshev theorem that x^* is a point of equioscillation. We denote it by t_2 and $\|R^*\|_\infty = |R^*(t_2)|$.

By uniqueness of $x^* = t_2 \in (a,b)$, other such points cannot be inside (a,b). Hence, both t_1 e t_3 are on the frontier, $t_1 = a$ and $t_3 = b$.

The system [2.1] becomes

$$\begin{cases} f(a) - (\theta_1 a + \theta_2) = \epsilon E^* \\ f(t_2) - (\theta_1 t_2 + \theta_2) = -\epsilon E^* \\ f'(t_2) = \theta_1 \\ f(b) - (\theta_1 b + \theta_2) = \epsilon E^* \end{cases} \qquad [2.2]$$

which yields θ_1, θ_2, $E^*$and t_2.

For example, let $f : [0,1] \to \mathbb{R}; x \mapsto f(x) := x^2$. Here $a = 0$, $b = 1$. In $x = 0$, R^* takes value $R^*(0) = E^*$; decreases until its minimal value $R^*(t_2) = -E^*$ and then increases until $R^*(1) = E^*$.

Therefore, [2.2] becomes

$$\begin{cases} (0)^2 - (\theta_1 0 + \theta_2) = E^* \\ t_2^2 - \theta_1 t_2 - \theta_2 = -E^* \\ \frac{d(R^*(x))}{dx} |_{x=t_2} = 0 \\ (1)^2 - \theta_1 1 - \theta_2 = E^* \end{cases}.$$

i.e.

$$\begin{cases} (\mathrm{a}) -\theta_2 = E^* \\ (\mathrm{b}) t_2^2 - \theta_1 t_2 - \theta_2 = -E^* \\ (\mathrm{c}) t_2 = \frac{\theta_1}{2} \\ (\mathrm{d}) 1 - \theta_1 - \theta_2 = E^* \end{cases}.$$

It follows that

$$\theta_1 = 1, \theta_2 = -\frac{1}{8}, t_2 = \frac{1}{2}, E^* = \frac{1}{8}.$$

The straight line which minimizes the norm $\left\|x^2 - (\theta_1 x + \theta_2)\right\|_\infty$ is thus

$$y = x - \frac{1}{8}$$

and

$$\left\|x^2 - (\theta_1 x + \theta_2)\right\|_\infty = \frac{1}{8} = E^*.$$

The set of equioscillation points of R^* is $\left\{0, \frac{1}{2}, 1\right\}$.

EXAMPLE 2.2.– Consider the case when

$$f : [0, 1] \to (0, +\infty), x \mapsto f(x) := e^x.$$

The best linear approximation of f is obtained solving

$$\begin{cases} (\mathrm{a}) 1 - \theta_2 = E^* \\ (\mathrm{b}) e^{t_2} - (\theta_1 t_2 + \theta_2) = -E^* \\ (\mathrm{c}) e^{t_2} = \theta_1 \\ (\mathrm{d}) e - (\theta_1 + \theta_2) = E^* \end{cases}.$$

This yields

$$1 - \theta_2 = e - (\theta_1 + \theta_2)$$

i.e. $\theta_1 = e - 1 \sim 1,7182$. Next $t_2 = \ln \theta_1 \sim 0,5413$. Also

$$\theta_1 - (0,5413\theta_1 + \theta_2) = -E^*$$

which, in turn, provides

$$E^* \sim 0,1059$$

which yields $\theta_2 = 1 - E^* \sim 0,8941$.

Frontier points of $[a,b]$ are not necessarily equioscillation points of R^*, in non-standard cases. Consider the case when f has more than one extremum in (a,b), hence f is neither convex nor concave. For example, let $f : [-2,1] \to \mathbb{R}, x \mapsto f(x) := x^3$, which does not fulfill the above hypotheses of theorem 2.1. Let $P^*(x) = \theta_1 x + \theta_2$ be the best uniform approximation of f on $[-1,1]$. Assume that $x_1 = -1$ and $x_2 = 1$ are equioscillation points of $R^*(x) := x^3 - P^*(x)$.

Then,

$$\begin{cases} (-1)^3 - (-\theta_1 + \theta_2) = \epsilon E^* \\ (1)^3 - (\theta_1 + \theta_2) = \epsilon E^* \\ 3t^2 = \theta_1 \\ t^3 - 3\theta_1 t - \theta_2 = -\epsilon E^* \end{cases}.$$

Hence, $\epsilon = 1$ and the system has solutions $t_{1,2} \sim \pm 0,5773$, $t_3 = -2, t_4 = 1$. Thus, we obtain the two minimax solutions

$$P^*(x) = x - \frac{\sqrt{3}}{9} \text{ and } P^*(x) = x + \frac{\sqrt{3}}{9}$$

which cannot hold due to uniqueness of the minimax polynomial.

It follows that two equioscillation points should be found therein. For example, the alternating set might be $\{u_1, u_2, 1\}$ with $(u_1, u_2) \in (-1,1) \times (-1,1)$. In this case

$$\begin{cases} u_1^3 - (\theta_1 u_1 + \theta_2) = E^* \\ u_2^3 - (\theta_1 u_2 + \theta_2) = -E^* \\ 1^3 - (\theta_1 1 + \theta_2) = E^* \end{cases}$$

which produces

$$P^*(x) = \frac{3}{4}x, \; \|R^*\|_\infty = \frac{1}{4}.$$

The choice $\{-1, u_1, u_2\}$ produces the same minimax solution.

If instead of a linear approximation we look for a quadratic approximation of the function x^3, then the number of equioscillation points should be $\deg(\theta_1 x^2 + \theta_2 x + \theta_3) + 2 = 4$. Both frontier points are equioscillation points when $f'''(x) \neq 0$ for all $x \in (-1, 1)$. Since $f'''(x) = 6$, the alternating points for R^* are of the form $\{-1, t_1, t_2, 1\}$.

More generally, when the minimax polynomial is to be found in $\mathcal{P}_n$ then the frontier points of $[a, b]$ are alternating points of R^* if the $(n+1)-th$ derivative $f^{(n+1)}$ takes non-null value on (a, b).

2.3. Convergence of the minimax approximation

It seems intuitively clear that the minimax approximation should converge as the degree of the approximating polynomial increases, on the basis of the Weierstrass theorem on continuous functions defined on a compact set.

As f is continuous on the compact set $[a, b]$, the Heine–Cantor theorem implies that f is uniformly continuous on $[a, b]$.

Hence, for all positive ε there exists some $n \in \mathbb{N}$ and some polynomial Q_n with degree n, such that the inequality

$$|f(x) - Q_n(x)| \leq \varepsilon$$

holds for all $x \in [a, b]$, which yields

$$\|f - Q_n\|_\infty = \sup_{x \in [a,b]} |f(x) - Q_n(x)| \leq \varepsilon.$$

Let P_n^* be the minimax polynomial with degree n. Then,

$$0 \leq \|f - P_n^*\|_\infty \leq \|f - Q_n\|_\infty .$$

Using the Weierstrass theorem

$$\|f - Q_n\|_\infty \to 0 \text{ as } n \to \infty.$$

Hence,

$$\|f - P_n^*\|_\infty \to 0 \text{ as } n \to \infty.$$

The above convergence of the minimax approximation may be very slow. To approximate the function $x \mapsto \arcsin x$ on $[-1, 1]$, with an error less than 10^{-6}, the degree of the minimax polynomial should exceed $n = 1000$.

The convergence may also be quite fast: if $f \in \mathcal{C}^{(\infty)}([a, b])$, then the sup norm goes to 0 faster than $n^{-\alpha}$ for all $\alpha \geq 1$; this remark is the starting point of the next section.

2.3.1. *Rate of convergence of the minimax approximation*

It holds

THEOREM 2.2.– Let $f \in \mathcal{C}^{(n+1)}([-1, 1])$, with $f^{(n+1)}(x) \neq 0$ for all x in $[-1, 1]$. Assume that for all x in $[-1, 1]$,

$$m \leq \left| f^{(n+1)}(x) \right| \leq M.$$

Then, denoting P^* the minimax approximation with degree n,

$$\frac{m}{2^n (n+1)!} \leq \| f - P^* \|_\infty \leq \frac{M}{2^n (n+1)!}.$$

PROOF.– Consider the roots $x_1, ..., x_{n+1}$ of the function $R^* := f - P^*$. It holds that

$$f(x_i) = P(x_i), i = 1, ..., n+1.$$

Hence, P^* is the Lagrange interpolation polynomial of f with nodes on the roots of the function R^*. The interpolation error is given by

$$R^*(x) = \frac{f^{(n+1)}(\xi)}{(n+1)!} \prod_{i=1}^{n+1} (x - x_i),$$

for some $\xi \in (-1, 1)$; see Chapter 3, formula [3.6].The polynomial

$$\prod_{i=1}^{n+1} (x - x_i)$$

belongs to $\mathcal{P}_{n+1}$ and the coefficient of x^{n+1} is 1; hence,

$$\left\| \prod_{i=1}^{n+1} (x - x_i) \right\|_\infty \geq \frac{1}{2^n} = \left\| \widetilde{T}_{n+1} \right\|_\infty$$

where, $\widetilde{T}_{n+1}(t) = T_{n+1}/2^n$ (see Chapter 3, formula [3.4]) and T_{n+1} is the Chebyshev polynomial of first kind with degree $n+1$. Among all polynomials with degree $n+1$ with coefficient of x^{n+1}equal to 1, it has minimal norm. It follows that

$$\begin{aligned}\|R^*\|_\infty &:= \max_{x\in[-1,1]} |R^*| = \max_{x\in[-1,1]} \left| \frac{f^{(n+1)}(\xi)}{(n+1)!} \prod_{i=1}^{n+1} (x - x_i) \right| \\ &= \frac{\max_{x\in[-1,1]} \left|f^{(n+1)}(x)\right|}{(n+1)!} \max_{x\in[-1,1]} \prod_{i=1}^{n+1} |x - x_i| \\ &= \frac{\max_{x\in[-1,1]} \left|f^{(n+1)}(x)\right|}{(n+1)!} \left\| \prod_{i=1}^{n+1} (x - x_i) \right\|_\infty \\ &\geq \frac{m}{(n+1)!} 2^{-n}\end{aligned}$$

since through hypothesis, $m \leq \left|f^{(n+1)}(x)\right|$ for all x. Hence,

$$\frac{m}{(n+1)!} 2^{-n} \leq \|R^*\|_\infty .$$

We will now prove that

$$\|R^*\|_\infty \leq \frac{M}{2^n (n+1)!}.$$

Let us consider, by contradiction,

$$\|R^*\|_\infty > \frac{M}{2^n (n+1)!}. \qquad [2.3]$$

From

$$R^*(x) = \frac{f^{(n+1)}(\xi)}{(n+1)!} \prod_{i=1}^{n+1} (x - x_i)$$

it follows that

$$\prod_{i=1}^{n+1} (x - x_i) = \frac{(n+1)! R^*(x)}{f^{(n+1)}(\xi)}.$$

Hence, up to a constant $K := (n+1)!/f^{(n+1)}(\xi)$ both polynomials $\prod_{i=1}^{n+1}(x-x_i)$ and $R^*(x)$ are equal. Hence,

$$\left\|\prod_{i=1}^{n+1}(x-x_i)\right\|_\infty = K\|R^*\|_\infty.$$

Multiply both sides in [2.3] by K, it follows that

$$K\|R^*\|_\infty > \frac{MK}{2^n(n+1)!},$$

i.e.

$$\left\|\prod_{i=1}^{n+1}(x-x_i)\right\|_\infty > \frac{MK}{2^n(n+1)!} = \frac{M\frac{(n+1)!}{f^{(n+1)}(\xi)}}{2^n(n+1)!}$$
$$= \frac{M}{2^n f^{(n+1)}(\xi)} > \frac{1}{2^n}$$

since $M \geq f^{(n+1)}(\xi)$ and, therefore,

$$\frac{M}{f^{(n+1)}(\xi)} \geq 1.$$

Hence, formula [2.3] implies

$$\left\|\prod_{i=1}^{n+1}(x-x_i)\right\|_\infty > \frac{1}{2^n}. \qquad [2.4]$$

Since,

$$\prod_{i=1}^{n+1}(x-x_i) = KR^*(x)$$

both the polynomials $\prod_{i=1}^{n+1}(x-x_i)$ and $R^*(x)$ have same $n+2$ equioscillations. The polynomial $\prod_{i=1}^{n+1}(x-x_i)$ has $n+2$ equioscillations, with changing signs, and with an amplitude of oscillation, larger in absolute value than 2^{-n} (see [2.4]). Define the polynomial now

$$Q(x) := \prod_{i=1}^{n+1}(x - x_i) - T_{n+1}(x)$$

where T_{n+1} denotes the Chebyshev polynomial of first kind, see Appendix 2.

It holds that $Q \in \mathcal{P}_n$. The polynomial Q oscillates $n + 2$ times (with no common values); if $\prod_{i=1}^{n+1}(x - x_i)$ equioscillates at points $y_1, ..., y_{n+2}$ with amplitude $\eta := \left\| \prod_{i=1}^{n+1}(x - x_i) \right\|_\infty$, then on those very y_i's,

$$\begin{aligned} Q(y_i) &= \prod_{j=1}^{n+1}(y_i - x_j) - T_{n+1}(y_i) \\ &= \epsilon_i \eta - T_{n+1}(y_i) := \delta_i. \end{aligned}$$

with $\epsilon_i := sgn\left(\prod_{i=1}^{n+1}(y_i - x_i)\right)$. Now $\delta_i > 0$, for all i, since

$$\left| \prod_{j=1}^{n+1}(y_i - x_j) \right| = \eta > 2^{-n} = \max_{x \in [-1,1]} |T_{n+1}(x)|.$$

The δ_is change sign from i to $i+1$ for all i. Hence, Q oscillates on the $n+2$ points $y_1, ..., y_{n+2}$; note that the y_i's are neither alternating nor equioscillation points for Q. However, the δ_is do not assume common values for different is. Clearly, $\delta_i \neq 0$ since $\prod_{j=1}^{n+1}(y_i - x_j) = T_{n+1}(y_i)$ never holds. Hence, Q oscillates in $n + 2$ points with an alternating sign. Thus, it crosses $n + 1$ times the axis of the abscissa. Hence, Q has $n + 1$ roots. Since $Q \in \mathcal{P}_n$, it follows that Q is the null polynomial, which cannot hold, and the proof is complete. □

2.4. Proof of the de la Vallée Poussin theorem

Before going to the proof of the convergence of the Remez algorithm, we will examine a result of de la Vallée Poussin. This theorem provides a lower bound for $R^* := \|f - P^*\|_\infty$, where P^* is the minimax polynomial for f in $\mathcal{P}_n$.

THEOREM 2.3 (de la Vallée Poussin).– Let $f \in \mathcal{C}^{(0)}([a,b])$ and $P_n \in \mathcal{P}_n$. Assume that there exist $n+2$ points $x_0, ..., x_{n+1}$, such that

$$(f - P)(x_i) = (-1)^i E_i \quad [2.5]$$

for $i = 0, ..., n+1$, where the constants E_i are all non-null and have the same sign. Then,

$$R^* \geq \min \{|E_0|, ..., |E_{n+1}|\}.$$

PROOF.– By contradiction, assume that there exists a polynomial Q in $\mathcal{P}_n$ and

$$\|f - Q\|_\infty < \min \{|E_0|, ..., |E_{n+1}|\}$$

where we assume that all E_i 's are positive. We obtain

$$|f(x_i) - Q(x_i)| \leq \|f - Q\|_\infty < \min \{|E_0|, ..., |E_{n+1}|\} \leq |E_i|$$

for all $i = 0, ..., n+1$. Now, stating that for some i

$$|f(x_i) - Q(x_i)| \leq |E_i| \quad [2.6]$$

amounts to saying that for such i, we may find some constant $k(i)$

$$k(i) := \frac{(sign(f(x_i) - Q(x_i)))(f(x_i) - Q(x_i))}{E_i}$$

such that

$$\begin{cases} f(x_0) - Q(x_0) = k(0)\ E_0 \\ \dots\dots\dots\dots\dots\dots\dots\dots \\ f(x_j) - Q(x_j) = k(j)\ E_j \\ \dots\dots\dots\dots\dots\dots\dots\dots \\ f(x_{n+1}) - Q(x_{n+1}) = k(i)\, E_i \end{cases}.$$

All constants $k(i)$ satisfy

$$-1 < k(i) < +1$$

since for all j

$$\frac{|f(x_j) - Q(x_j)|}{E_j} = |k(j)|;$$

By hypothesis

$$\frac{|f(x_j) - Q(x_j)|}{E_j} < 1$$

it holds for all $j = 0, ..n + 1$, that

$$|k(j)| < 1.$$

Now, denote

$$T := P - Q$$

where P satisfies [2.5]. Evaluate T on the x_i's. This yields, after some calculus

$$sgn(T(x_i)) = sgn\left((-1)^{i+1}\right).$$

Using $sgn(d_i) = +1$

$$\begin{aligned} sgn(T(x_i)) &= sgn\left((-1)^{i+1}\right) = sgn\left(\left((-1)^{i+1} + k(i)\right)E_i\right) \\ &= sgn\left((-1)^{i+1} + k(i)\right) sgn(E_i) = sgn\left((-1)^{i+1} + k(i)\right) \end{aligned}$$

which, along with

$$-1 < k(i) < 1$$

yields

$$(-1)^{i+1} - 1 < (-1)^{i+1} + k(i) < 1 + (-1)^{i+1}. \qquad [2.7]$$

For $i = 0$, [2.7] becomes

$$-2 < (-1)^{0+1} + k(0) < 0;$$

hence,

$$sgn\left((-1)^{i+1} + k(i)\right) = -1 = sgn\left((-1)^{i+1}\right).$$

For $i = 1$, [2.7] becomes

$$0 < 1 + k(0) < 2$$

hence,

$$sgn\left((-1)^{i+1}+k(i)\right)=1=sgn\left((-1)^{i+1}\right).$$

and so on. We, thus, have found a polynomial

$$T:=P-Q$$

in $\mathcal{P}_n$ which alternates sign $n+1$ times, and therefore, with $n+1$ roots. This contradiction achieves the proof. □

EXAMPLE 2.3.– Let

$$f:[-\pi,\pi]\to\mathbb{R}\,;x\mapsto f(x)=\sin x.$$

Consider the points

$$x_0=-\pi,\,x_1=-\frac{\pi}{2},\,x_2=\frac{\pi}{2},\,x_3=\pi.$$

We evaluate an lower bound of

$$E^*:=\left\|f-\left(\frac{3}{4\pi}x-\frac{1}{4}\right)\right\|_\infty.$$

It is enough to calculate the minimal value of

$$E_i:=E(x_i):=f(x_i)-\left(\frac{3}{4\pi}x_i-\frac{1}{4}\right).$$

It holds $E_0=\sin(-\pi)-\left(\left(\frac{3}{4\pi}(-\pi)-\frac{1}{4}\right)\right)=1$, $E_1=-\frac{3}{8}$, $E_2=\frac{7}{8}$, $E_3=-\frac{1}{2}$. Hence,

$$\left\|f-\left(\frac{3}{4\pi}x-\frac{1}{4}\right)\right\|_\infty>\min\left\{|1|,\left|-\frac{3}{8}\right|,\left|\frac{7}{8}\right|,\left|-\frac{1}{2}\right|\right\}=\frac{3}{8}.$$

The Borel–Chebyshev theorem does not provide any explicit construction for the minimax polynomial. Numerical solutions do exist. Such is the scope of the following section.

2.5. The Yevgeny Yakovlevich Remez algorithm

We begin with a simple and important lemma.

LEMMA 2.1.– Let $f \in \mathcal{C}^{(0)}([a,b])$, and P_n be a polynomial in $\mathcal{P}_n$. Let $R := f - P_n$. There exists such P_n for which the function R equioscillates on $n+2$ distinct points

$$a \leq y_1 < ... < y_{n+2} \leq b$$

in $[a,b]$. The points y_i's can be chosen arbitrarily. An adequate choice of those points allows to define P_n with exact degree n.

PROOF.– Choose in $[a,b]$, $n+2$ points $y_1, ..., y_{n+2}$, pairwise distinct. The equioscillation property means that $R(y_i) = (-1)^i E$. Hence,

$$P_n(y_i) = f(y_i) - (-1)^i E.$$

Since, $P_n \in \mathcal{P}_n$

$$P(x) = f(x) - (-1)^i E$$

cannot hold on any arbitrary x. Consider the Lagrange interpolation polynomial Q_{n+1} in $\mathcal{P}_{n+1}$, which interpolates f on the points y_i, and the Lagrange interpolation polynomial S_{n+1} in $\mathcal{P}_{n+1}$, which interpolates the function $(-1)^i$ on the same points y_i, denote

$$\begin{cases} Q_{n+1}(x) = q_0 + q_1 x + ... + q_{n+1} x^{n+1} \\ S_{n+1}(x) = s_0 + s_1 x + ... + s_{n+1} x^{n+1} \end{cases}$$

and solve the linear system

$$\begin{cases} Q_{n+1}(y_1) = q_0 + q_1 y_1 + ... + q_{n+1} y_1^{n+1} \\ \dots\dots\dots\dots\dots\dots\dots\dots\dots\dots \\ Q_{n+1}(y_{n+1}) = q_0 + q_1 y_{n+1} + ... + q_{n+1} y_{n+1}^{n+1} \\ S_{n+1}(y_1) = s_0 + s_1 y_1 + ... + s_{n+1} y_1^{n+1} \\ \dots\dots\dots\dots\dots\dots\dots\dots\dots\dots \\ S_{n+1}(y_{n+1}) = s_0 + s_1 y_{n+1} + ... + s_{n+1} y_{n+1}^{n+1} \end{cases}.$$

Let $Q_{n+1}(x) = \overline{q}_0 + \overline{q}_1 x + ... + \overline{q}_{n+1} x^{n+1}$ and $S_{n+1}(x) = \overline{s}_0 + \overline{s}_1 x + ... + \overline{s}_{n+1} x^{n+1}$ be the resulting polynomials. That the degree of Q_{n+1} and of S_{n+1} may be exactly n as an adequate tuning of $y_1, .., y_{n+1}$ is clear. Denote

$$E := \frac{\overline{q}_{n+1}}{\overline{s}_{n+1}} \text{ and } P_n := Q_{n+1} - S_{n+1} E.$$

It holds by construction that

$$\begin{aligned}P_n(y_i) &:= Q_{n+1}(y_i) - S_{n+1}(y_i)E \\ &= \left(\overline{q}_0 + \overline{q}_1 y_i + ... + \overline{q}_{n+1} y_i^{n+1}\right) - \left(\overline{s}_0 + \overline{s}_1 y_i + ... + \overline{s}_{n+1} y_i^{n+1}\right) \frac{\overline{q}_{n+1}}{\overline{s}_{n+1}} \\ &= \frac{1}{\overline{s}_{n+1}} \left(\overline{q}_0 \overline{s}_{n+1} + \overline{q}_1 \overline{s}_{n+1} y_i + ... + \overline{q}_n \overline{s}_n y_i^n - \overline{s}_0 \overline{q}_{n+1} - \overline{s}_1 y_i - ... - \overline{s}_n y_i^n\right) \\ &= \frac{\overline{q}_0 \overline{s}_{n+1} - \overline{s}_0 \overline{q}_{n+1}}{\overline{s}_{n+1}} + \left(\frac{\overline{q}_1 \overline{s}_{n+1} - \overline{s}_1}{\overline{s}_{n+1}}\right) y_i + ... + \left(\frac{\overline{q}_n \overline{s}_n - \overline{s}_n}{\overline{s}_{n+1}}\right) y_i^n\end{aligned}$$

and therefore,

$$P_n(x) = \frac{\overline{q}_0 \overline{s}_{n+1} - \overline{s}_0 \overline{q}_{n+1}}{\overline{s}_{n+1}} + \left(\frac{\overline{q}_1 \overline{s}_{n+1} - \overline{s}_1}{\overline{s}_{n+1}}\right) x + ... + \left(\frac{\overline{q}_n \overline{s}_n - \overline{s}_n}{\overline{s}_{n+1}}\right) x^n.$$

We see that $P_n \in \mathcal{P}_n$, that when $x = y_i$, then $R(y_i) = (-1)^i E$ with $E = \frac{\overline{q}_{n+1}}{\overline{s}_{n+1}}$. Indeed $P_n(y_i) := Q_{n+1}(y_i) - S_{n+1}(y_i)E = f(x_i) - (-1)^i E$. Hence P_n solves the problem raised in the lemma; and deg P_n equals n. □

REMARK 2.1.– When the y_i's are the points of maximal and minimal value of R then P_n is the minimax polynomial of f. This remark provides the clue for the following algorithm.

2.5.1. *The Remez algorithm*

1) Let us consider the function f is the input, and $n+2$ pairwise distinct points in $[a,b]$, $a \leq y_1(\mathbf{1}) < ... < y_{n+2}(\mathbf{1}) \leq b$. With these points we evaluate the two polynomials with degree $n+1$, $\overline{Q}_{n+1}^{(\mathbf{1})}$ and $\overline{S}_{n+1}^{(\mathbf{1})}$, which are defined in lemma 2.1, solving in $\left(\overline{q}_0^{(\mathbf{1})}, \overline{q}_1^{(\mathbf{1})}, ..., \overline{q}_{n+1}^{(\mathbf{1})}\right)$ and $\left(\overline{s}_0^{(\mathbf{1})}, \overline{s}_1^{(\mathbf{1})}, ..., \overline{s}_{n+1}^{(\mathbf{1})}\right)$ the system

$$\begin{cases} f(y_i(\mathbf{1})) = q_0^{(\mathbf{1})} + q_1^{(\mathbf{1})} x_i + ... + q_{n+1}^{(\mathbf{1})} x_i^{n+1}, \ i = 1, ..., n+2 \\ (-1)^i E^{(\mathbf{1})} = \overline{s}_0^{(\mathbf{1})} + \overline{s}_1^{(\mathbf{1})} x_i + ... + \overline{s}_{n+1}^{(\mathbf{1})} x_i^{n+1}, \ i = 1, ..., n+2 \end{cases}$$

with

$$E^{(\mathbf{1})} := \frac{\overline{q}_{n+1}^{(\mathbf{1})}}{\overline{s}_{n+1}^{(\mathbf{1})}}.$$

and it holds that

$$\overline{P}_n^{(\mathbf{1})} := \overline{Q}_{n+1}^{(\mathbf{1})} - \overline{S}_{n+1}^{(\mathbf{1})} E^{(\mathbf{1})}. \qquad [2.8]$$

Since $\left|\overline{S}_{n+1}^{(1)}(y_i(\mathbf{1}))\right| = 1$, for all $i = 1, ..., n+2$, it results from [2.8] that

$$\left|E^{(1)}\right| = \left|\overline{P}_n^{(1)}(y_i(\mathbf{1})) - \overline{Q}_{n+1}^{(1)}(y_i(\mathbf{1}))\right|.$$

Clearly, $\overline{P}_n^{(1)}$ belongs to $\mathcal{P}_n$. We, thus, have to choose the initial $y_1(\mathbf{1}), ..., y_{n+2}(\mathbf{1})$ such that both $\overline{Q}_{n+1}^{(1)}$ and $\overline{S}_{n+1}^{(1)}$ have exact degree $n+1$, along with

$$a \leq y_1(\mathbf{1}) < ... < y_{n+2}(\mathbf{1}) \leq b.$$

The error function pertaining to $\overline{P}_n^{(1)}$ is,

$$R_1 := f - \overline{P}_n^{(1)}$$

If the points, $y_1(\mathbf{1}), ..., y_{n+2}(\mathbf{1})$, are all points of maximal and minimal value of R_1, then

$$\overline{P}_n^{(1)} = P^*$$

where we indicate by P^* the minimax polynomial of f, then stop.

Otherwise, it holds that $\overline{P}_n^{(1)} \neq P^*$, then we go to the next step.

2) In this step we update the equioscillation points , using a nearest neighbor procedure, as follows:

For z, a point of maximal or minimal value of R_1 and $z \in (y_i(\mathbf{1}), y_{i+1}(\mathbf{1}))$ with $sgn((R_1)(z)) = R_1(y_i(\mathbf{1}))$, we substitute $y_i(\mathbf{1})$ by z. If, instead, $sgn(R_1(z)) = R_1(y_{i+1}(\mathbf{1}))$ then, we substitute $y_{i+1}(\mathbf{1})$ by z.

It may occur that in $(y_i(\mathbf{1}), y_{i+1}(\mathbf{1}))$ no point of maximal or minimal value of R_1 exists; we keep both points $y_i(\mathbf{1}), y_{i+1}(\mathbf{1})$ in the updated list in such cases.

It may also occur that $(y_i(\mathbf{1}), y_{i+1}(\mathbf{1}))$ contains many points of extreme value of R_1. For example, assume two minimal values at z_1, z_2 and three maximal values at u_1, u_2, u_3. Then, we consider only the point $z := \min\{z_1, z_2\}$ and the point $u := \max\{u_1, u_2, u_3\}$. Hence, if z satisfies $sgn(R_1(z)) = R_1(y_i(\mathbf{1}))$ or $sgn(R_1(z)) = R_1(y_{i+1}(\mathbf{1}))$ we substitute $y_i(\mathbf{1})$ or $y_{i+1}(\mathbf{1})$ by z. In the same way, if u is such that $sgn(R_1(u)) = R_1(y_i(\mathbf{1}))$ or $sgn(R_1(u)) = R_1(y_{i+1}(\mathbf{1}))$, we substitute $y_i(\mathbf{1})$ or $y_{i+1}(1)$ by u.

Those substitutions yield a new sequence of points

$$a \leq y_1(\mathbf{2}) < ... < y_{n+2}(\mathbf{2}) \leq b$$

which are considered as new equioscillation points, which are either all of them, or some of them, or none, points of extreme values for $R_{\mathbf{1}}$.

To identify this property, we use the first order condition, solving

$$\frac{dR_{\mathbf{1}}(x)}{dx} = 0.$$

when

$$\frac{d^{n+2}R_{\mathbf{1}}(x)}{dx^{n+2}} > 0$$

for all $x \in [a, b]$, then the function R_1 is standard which ensures that

$$y_1(\mathbf{k}) = a \text{ and } y_{n+2}(\mathbf{k}) = b.$$

This allows us to use the same procedure as in **1** and to identify a new polynomial $\overline{P}_n^{(\mathbf{2})}$ along with a new value for $E^{(\mathbf{2})}$.

If all points $y_1(\mathbf{2}) < ... < y_{n+2}(\mathbf{2})$ are maximum/minimum points of $R_{\mathbf{1}}$, then $\overline{P}_n^{(2)} = P^*$, stop.

Otherwise we turn to phase **3.**

3) We define a new set of points $a \leq y_1(\mathbf{3}) < ... < y_{n+2}(\mathbf{3}) \leq b$, from which a new polynomial, $\overline{P}_n^{(3)}$, with amplitude of oscillation $E^{(3)}$, as in step **2.**

The points $y_1(\mathbf{3}), ..., y_{n+2}(\mathbf{3})$ are obtained through the comparison with the extreme values of R_2, as previously, and substituting $y_1(\mathbf{2}), ..., y_{n+2}(\mathbf{2})$ respecting the alternating signs of $R_2(y_i(\mathbf{2}))$. Also if,

$$\frac{d^{n+2}R_{\mathbf{1}}(x)}{dx^{n+2}} > 0$$

for all $x \in [a, b]$, then

$$a = y_1(\mathbf{1}) = y_1(\mathbf{2}) = y_1(\mathbf{3}) \text{ and } b = y_{n+2}(\mathbf{1}) = y_{n+2}(\mathbf{2}) = y_{n+2}(\mathbf{3}).$$

Proceed from step **k** to step **k+1** as from step **2** to step **3**.

We should stop when we obtain $\overline{P}_n^{(\mathbf{k})}$, such that

$$\left|P^* - \overline{P}_n^{(\mathbf{k})}\right| \leq \varepsilon$$

with ε chosen as an accuracy bound.

The actual stopping rule should now be examined.

At step **k,** we selected the set of equioscillation points

$$\{y_1(\mathbf{k}), \ldots, y_{n+2}(\mathbf{k})\}$$

for $R_{\mathbf{k}}$ where, if

$$\frac{d^{n+2}R_1(x)}{dx^{n+2}} > 0$$

for all $x \in [a, b]$, we have chosen

$$y_1(\mathbf{k}) = a, \ y_{n+2}(\mathbf{k}) = b.$$

On the local extrema $z_i(\mathbf{k})$ of $R_{\mathbf{k}}$, namely for points such that

$$\frac{dR_{\mathbf{k}}(x)}{dx}\,|_{x=z_i(\mathbf{k})} = 0$$

we have

$$|R_{\mathbf{k}}(z_i(\mathbf{k}))| \geq |R_{\mathbf{k}}(y_i(\mathbf{k}))|.$$

Either $y_i(\mathbf{k})$ is an equioscillation point for $R_{\mathbf{k}}$ or it is a local extremum of $R_{\mathbf{k}}$, hence

$$|R_{\mathbf{k}}(z_i(\mathbf{k}))| = |R_{\mathbf{k}}(y_i(\mathbf{k}))|$$

or

$$|R_{\mathbf{k}}(z_i(\mathbf{k}))| > |R_{\mathbf{k}}(y_i(\mathbf{k}))|$$

since, $z_i(\mathbf{k})$ is a local extremum of $R_{\mathbf{k}}$.

Since $y_i(\mathbf{k})$ is an equioscillation point of $R_{\mathbf{k}}$

$$|R_k(y_i(\mathbf{k}))| = \left| f(y_i(\mathbf{k})) - \overline{P}_n^{(\mathbf{k})}(y_i(\mathbf{k})) \right|$$

$$= \left| (-1)^i E^{(\mathbf{k})} \right| = \left| E^{(\mathbf{k})} \right| = \frac{\overline{q}_{n+1}^{(\mathbf{k})}}{\overline{s}_{n+1}^{(\mathbf{k})}}.$$

Hence,

$$|R_{\mathbf{k}}(z_i(\mathbf{k}))| \geq \frac{\overline{q}_{n+1}^{(\mathbf{k})}}{\overline{s}_{n+1}^{(\mathbf{k})}}$$

for $i = 1, ..., n+1$.

The stopping rule is, then, as follows.

Denote that

$$M^{(\mathbf{k})} := \max\left\{|R_{\mathbf{k}}(y_1(\mathbf{k}))|, ..., |R_{\mathbf{k}}(y_{n+2}(\mathbf{k}))|\right\}$$

$$m^{(\mathbf{k})} := \min\left\{|R_{\mathbf{k}}(y_1(\mathbf{k}))|, ..., |R_{\mathbf{k}}(y_{n+2}(\mathbf{k}))|\right\}.$$

If

$$\left| \frac{M^{(\mathbf{k})}}{m^{(\mathbf{k})}} - 1 \right| < \varepsilon \qquad [2.9]$$

with $\varepsilon > 0$ fixed, then the polynomial $\overline{P}_n^{(\mathbf{k})}$ is the solution.

If [2.9] holds, then the oscillation of $R_{\mathbf{k}}$ is nearly the same on all $[a, b]$ and $R_{\mathbf{k}}$ equioscillates on its local extrema alternating sign.

2.5.2. *Convergence of the Remez algorithm*

2.5.2.1. *Some technical results*

Denote that

$$V := \begin{pmatrix} x_0^0 & . & x_0^j & . & x_0^n \\ . & . & . & . & . \\ x_i^0 & . & x_i^j & . & x_i^n \\ . & . & . & . & . \\ x_n^0 & . & x_n^j & . & x_n^n \end{pmatrix}$$

LEMMA 2.2.– It holds that $\det V = \prod_{i=0,...,n\,;\,j=0,...,n+1\,;\,j>i} (x_j - x_i)$. The matrix V^{-1} exists, if and only if, the points $x_0, ..., x_{n+1}$ are pairwise distinct.

PROOF.– It is enough to apply the Laplace theorem after the following transformations

$$\begin{pmatrix} x_0^j \\ . \\ x_i^j \\ . \\ x_n^j \end{pmatrix} \curvearrowright \begin{pmatrix} x_0^j - x_n x_0^{j-1} \\ . \\ x_i^j - x_n x_i^{j-1} \\ . \\ x_n^j - x_n x_n^{j-1} \end{pmatrix} = \begin{pmatrix} x_0^j - x_n x_0^{j-1} \\ . \\ x_i^j - x_n x_i^{j-1} \\ . \\ 0 \end{pmatrix}$$

for $j = 2, ..., n+1$.

LEMMA 2.3.– Let $f \in \mathcal{C}^{(0)}([a,b])$. Consider $n+2$ points $x_0, ..., x_{n+1}$, pairwise distinct in $[a,b]$. Then, the system

$$\begin{cases} \sum_{i=0}^{n+1} x_i^j \alpha_i = 0 \text{for } j = 0, ..., n \\ \sum_{i=0}^{n+1} |\alpha_i| = 1 \text{ with } \alpha_0 > 0 \end{cases} \qquad [2.10]$$

has a unique solution $\alpha_i = (-1)^i |\alpha_i|$ with $i = 0, ..., n+1$. Furthermore, the system in the variables β_i's and E

$$\sum_{i=0}^{n} x_i^j \beta_i + (-1)^i E = f(x_i) \qquad [2.11]$$

with $i = 0, ..., n+1$ has a unique solution and

$$E = \sum_{i=0}^{n+1} f(x_i) = \sum_{i=0}^{n+1} \alpha_i (f(x_i) - T(x_i))$$

whatever the polynomial $T \in \mathcal{P}_n$.

PROOF.– The system $\sum_{i=0}^{n+1} x_i^j \alpha_i = 0$, $j = 0, ..., n$, has an infinite number of non-null solutions which depend upon a common multiplicative factor. Its matrix M has dimension $(n+1) \times (n+2)$ and maximal range, since its minors of order $n \times 1$ are transposed matrices of Vandermonde matrices built on distinct points. Let α_0 be the common multiplicative factor; assuming $\alpha_0 > 0$ implies from the fact that $\sum_{i=0}^{n+1} |\alpha_i| = 1$ that the $(n+2) - uple$ $(\alpha_0, ..., \alpha_{n+1})$ is uniquely determined. Hence, the system [2.10] has a unique solution.

Writing $\sum_{i=0}^{n+1} x_i^j \alpha_i = 0$ as $\sum_{i=0}^{n+1} x_i^j \alpha_i = -x_0^j \alpha_0$, for $j = 0, ..., n$ and applying Cramer theorem, together with [2.10] we obtain $\alpha_i = -\alpha_0 l_i(x_0)$, $i = 0, .., n+1$. The elementary Lagrange polynomial

$$l_i(x_0) := \prod_{j=1,\, j \neq i}^{n} \frac{x_0 - x_j}{x_i - x_j}$$

is a product of $2n - i - 1$ negative factors, hence α_i and α_{i+1} have opposite signs. We prove that for any polynomial $T \in \mathcal{P}_n$, it holds $\sum_{i=0}^{n+1} \alpha_i T(x_i) = 0$. Since from [2.10], $\sum_{j=0}^{n+1} \alpha_i x_i^j = 0$ for all x_i this follows linearity. System [2.11] has a unique solution, since the last column of its matrix has terms alternating values $+1$ and -1. Developing the determinant with respect to this last column, the module of this determinant equals the sum of determinants of Vandermonde matrices, all positive, since $x_j > x_i$ for $i > j$. Therefore, the matrix of the system [2.11] is invertible. Denote that $Q(x) := \sum_{j=0}^{n} \beta_j x^j$; the system [2.11] may be written as

$$f(x_i) - Q(x_i) = (-1)^i E$$

for all i. Multiply by α_i on both sides and sum upon i to obtain

$$\sum_{i=0}^{n+1} \alpha_i (f(x_i) - Q(x_i)) = \sum_{i=0}^{n+1} \alpha_i (-1)^i E = E \sum_{i=0}^{n+1} |\alpha_i| .$$

From the second line in [2.10], it holds that $\sum_{i=0}^{n+1} |\alpha_i| = 1$. Hence,

$$E = \sum_{i=0}^{n+1} \alpha_i (f(x_i) - Q(x_i)) . \qquad [2.12]$$

Now, for any $T \in \mathcal{P}_n$ it holds that $\sum_{i=0}^{n+1} \alpha_i T(x_i) = 0$. Hence, $\sum_{i=0}^{n+1} \alpha_i Q(x_i) = 0$ and, therefore, we may substitute Q by any polynomial $T \in \mathcal{P}_n$ in [2.12]. □

An important point in the Remez algorithm lies in the fact that for all **k**, all points $\mathbf{y}(\mathbf{k}) := (y_1(\mathbf{k}), \ldots, y_{n+2}(\mathbf{k}))$ have distinct coordinates; this implies that the polynomial $\overline{P}_n^{(\mathbf{k})}$ has a constant degree n. We consider this problem through a description of the amplitude of the oscillation $|E^{(\mathbf{k})}|$. If this oscillation is positive for all **k**, then $dist(y_i(\mathbf{k}), y_{i+1}(\mathbf{k})) \neq 0$ for all $i = 1, ..., n+2$, and for all $\mathbf{k} \in \mathbb{N}$.

We start with the following remark.

REMARK 2.2.– Since the limit of the sequences $d_{\mathbf{k}}$:=$\{|y_i(\mathbf{k}) - y_{i+1}(\mathbf{k})|, \mathbf{k} \in \mathbb{N}\}$, for $i = 1, ..., n+2$ might not be defined, we consider infimum limits. From $d_{\mathbf{k}}$ we build a non-decreasing sequence which we will consider for large **k.** Define

$$i_1 := \inf\{dist(y_i(\mathbf{1}), y_{i+1}(\mathbf{1})), dist(y_i(\mathbf{2}), y_{i+1}(\mathbf{2})), \\ ..., dist(y_i(\mathbf{k}), y_{i+1}(\mathbf{k})), ... : \mathbf{k} \in \mathbb{N}\}$$

$$i_2 := \inf\{dist(y_i(\mathbf{2}), y_{i+1}(\mathbf{2})), dist(y_i(\mathbf{3}), y_{i+1}(\mathbf{3})), \\ ..., dist(y_i(\mathbf{k}), y_{i+1}(\mathbf{k})), ... : \mathbf{k} \in \mathbb{N}\},$$

$$i_3 := \inf\{dist(y_i(\mathbf{3}), y_{i+1}(\mathbf{3})), dist(y_i(\mathbf{4}), y_{i+1}(\mathbf{4})), \\ ..., dist(y_i(\mathbf{k}), y_{i+1}(\mathbf{k})), ... : \mathbf{k} \in \mathbb{N}\}, ...$$

REMARK 2.3.– Clearly, $i_1 \leq ... \leq i_k \leq ....$, since for any $(y_i(\mathbf{k}), y_{i+1}(\mathbf{k})) \in [a, b]^2$ it holds $dist(y_i(\mathbf{k}), y_{i+1}(\mathbf{k})) \leq b - a$, and $\{i_k : k \in \mathbb{N}\}$ is bounded. Hence, the sequence $\{i_k : k \in \mathbb{N}\}$ converges to some $l := \min\lim_{k\to\infty} d_{\mathbf{k}}$.

LEMMA 2.4.– Let $f \in \mathcal{C}^{(0)}([a, b])$ and $\mathbf{y}(\mathbf{k})$ defined as above. If for all $\mathbf{k}$, $\left|E^{(\mathbf{k})}\right| \geq \gamma > 0$ holds, then for all $\mathbf{k} \in \mathbb{N}$ and for any $i = 1, ..., n+2$, the points $y_1(\mathbf{k}), ..., y_{n+2}(\mathbf{k})$ are distinct. It then holds that

$$dist(y_i(\mathbf{k}), y_{i+1}(\mathbf{k})) \geq \gamma,$$

where γ does not depend on **k**.

PROOF.– The proof is obtained through contradiction. Assume that

$$\min\lim_{k\to\infty} dist(y_i(\mathbf{k}), y_{i+1}(\mathbf{k})) = 0.$$

Define a polynomial $\widetilde{Q}_n$ in $\mathcal{P}_n$ which coincides with f on the $n+1$ following points

$$y_0(\mathbf{k}), ..., y_{j-1}(\mathbf{k}), y_{j+1}(\mathbf{k}), ..., y_{n+1}(\mathbf{k}).$$

From [2.12]

$$\left|E^{(\mathbf{k})}\right| = \left|\sum_{i=0}^{n+1} \alpha_i f(y_i(\mathbf{k})) - \widetilde{Q}_n(y_i(\mathbf{k}))\right|$$

$$= \left|\alpha_j \left(f(y_i(\mathbf{k})) - \widetilde{Q}_n(y_i(\mathbf{k}))\right)\right|$$

$$\leq |\alpha_j| \left(|f(y_i(\mathbf{k})) - f(y_{i+1}(\mathbf{k}))| + \left(\left|\widetilde{Q}_n(y_{i+1}(\mathbf{k})) - \widetilde{Q}_n(y_i(\mathbf{k}))\right|\right)\right).$$

Since through hypothesis

$$\min \lim_{k\to\infty} dist\left(y_i\left(\mathbf{k}\right), y_{i+1}\left(\mathbf{k}\right)\right) = 0$$

it follows that for any $\varepsilon > 0$, there exists k such that when $\mathbf{k} > k$,

$$\begin{cases} \left|f\left(y_i\left(\mathbf{k}\right)\right) - f\left(y_{i+1}\left(\mathbf{k}\right)\right)\right| \leq \varepsilon \\ \left|\widetilde{Q}_n\left(y_{i+1}\left(\mathbf{k}\right)\right) - \widetilde{Q}_n\left(y_i\left(\mathbf{k}\right)\right)\right| \leq \varepsilon \end{cases}.$$

Furthermore, $|\alpha_i| < 1$ and thus

$$0 < \gamma < \left|E^{(\mathbf{k})}\right| \leq |\alpha_j|\left(\varepsilon + \varepsilon\right) = 2\varepsilon\left|\alpha_j\right|$$

which cannot hold as ε is arbitrary. Hence,

$$\min \lim_{k\to\infty} dist\left(y_i\left(\mathbf{k}\right), y_{i+1}\left(\mathbf{k}\right)\right) \neq 0$$

and therefore,

$$dist\left(y_i\left(\mathbf{k}\right), y_{i+1}\left(\mathbf{k}\right)\right) \geq \gamma$$

with γ independent on $\mathbf{k}$.

THEOREM 2.4.– The sequence $\left\{\left|E^{(\mathbf{k+1})}\right| : \mathbf{k} \in \mathbb{N}\right\}$ is non-decreasing. Also there exists $\varepsilon > 0$ and some $\mathbf{k}_0$ such that for all $\mathbf{k} > \mathbf{k}_0$

$$E^* \leq \|R_{\mathbf{k}}\|_\infty \leq E^* + \varepsilon.$$

PROOF.– The Remez algorithm updates $y_i\left(\mathbf{k}\right)$ in $y_i^{(\mathbf{k+1})}$ as follows

$$y_i\left(\mathbf{k}\right) \curvearrowright y_i\left(\mathbf{k+1}\right)$$

$i = 0, ..., n+1$. Hence, we have

$$E^{(\mathbf{k+1})} = \sum_{i=0}^{n+1} \alpha_i \left(f\left(y_i\left(\mathbf{k+1}\right)\right) - P_n^{(\mathbf{k})}\left(y_i\left(\mathbf{k+1}\right)\right)\right) = \sum_{i=0}^{n+1} \alpha_i R_{\mathbf{k}}\left(y_i\right)$$

$$=_{(1)} \sum_{i=0}^{n+1} \alpha_i \left(\left|R_{\mathbf{k}}\left(y_i\left(\mathbf{k+1}\right)\right)\right| sgn\left(R_{\mathbf{k}}\left(y_i\left(\mathbf{k+1}\right)\right)\right)\right) =_{(2)} \sum_{i=0}^{n+1} |\alpha_i| \left|R_{\mathbf{k}}\left(y_i\left(\mathbf{k+1}\right)\right)\right|.$$

Equality (1) holds since $R_{\mathbf{k}}\left(y_i\left(\mathbf{k+1}\right)\right) = \left|R_{\mathbf{k}}\left(y_i\left(\mathbf{k+1}\right)\right)\right| sgn\left(R_{\mathbf{k}}\left(y_i\left(\mathbf{k+1}\right)\right)\right)$, (2) from the fact that α_i and $R_{\mathbf{k}}\left(y_i\left(\mathbf{k+1}\right)\right)$ have equal sign with

$$\alpha_i = (-1)^i \left|\alpha_i\right|.$$

From

$$E^{(\mathbf{k+1})} = \sum_{i=0}^{n+1} \left|\alpha_i\right| \left|R_{\mathbf{k}}\left(y_i\left(\mathbf{k+1}\right)\right)\right|$$

we obtain

$$E^{(\mathbf{k+1})} = \left|E^{(\mathbf{k+1})}\right|,$$

hence,

$$\left|E^{(\mathbf{k+1})}\right| = \sum_{i=0}^{n+1} \left|\alpha_i\right| \left|R_{\mathbf{k}}\left(y_i\left(\mathbf{k+1}\right)\right)\right|.$$

For any $i = 0, ..., n+1$, the point $y_i\left(\mathbf{k+1}\right)$ belongs to $\left[y_i\left(\mathbf{k}\right), y_i\left(\mathbf{k+1}\right)\right]$ and is a local extremum of $R_{\mathbf{k}}$. Hence, for all $y_i^{(\mathbf{k})}$, $i = 0, ..., n+1$,

$$R_{\mathbf{k}}\left(y_i\left(\mathbf{k+1}\right)\right) \geq R_{\mathbf{k}}\left(y_i\left(\mathbf{k}\right)\right) = E^{(\mathbf{k})};$$

hence,

$$\left|R_{\mathbf{k}}\left(y_i\left(\mathbf{k+1}\right)\right)\right| \geq \left|E^{(\mathbf{k})}\right|.$$

Now, $\left|R_{\mathbf{k}}\left(y_i\left(\mathbf{k+1}\right)\right)\right|$ is the amplitude of the oscillation of $P_n^{(\mathbf{k+1})}$, i.e. $R_{\mathbf{k}}\left(y_i\left(\mathbf{k+1}\right)\right) := E^{(\mathbf{k+1})}$ and thus

$$\left|E^{(\mathbf{k+1})}\right| \geq \left|E^{(\mathbf{k})}\right|.$$

Denoting

$$E^{(\mathbf{0})} := E := \frac{\overline{q}_{n+1}}{\overline{s}_{n+1}}$$

it holds that

$$\left|E^{(\mathbf{k})}\right| \geq \left|E^{(\mathbf{0})}\right|.$$

That $\left|E^{(\mathbf{0})}\right| > 0$ is a consequence of $\overline{q}_{n+1} > 0$ and $\overline{s}_{n+1} > 0$ (since Q and S have degree $n+1$).We claim that for all $\mathbf{k}$

$$\left|E^{(\mathbf{k})}\right| \leq E^*.$$

is the amplitude of the oscillation of $R_{\mathbf{k}} := f - P_n^{(\mathbf{k})}$. Hence,

$$E^* \geq \left|E^{(\mathbf{k})}\right| \geq \left|E^{(\mathbf{0})}\right| > 0.$$

Therefore, $\left\{E^{(\mathbf{k})} : \mathbf{k} \in \mathbb{N}\right\}$ is non-decreasing and upper bounded by E^*. Let

$$\eta\left(\mathbf{k}\right) := \left|E^{(\mathbf{k+1})}\right| - \left|E^{(\mathbf{k})}\right|$$

and consider the sequence $\{\eta\left(\mathbf{k}\right) : \mathbf{k} \in \mathbb{N}\}$. We have

$$\lim_{\mathbf{k}\to 0} \eta\left(\mathbf{k}\right) = 0.$$

Since $\left|E^{(\mathbf{k+1})}\right| = \sum_{i=0}^{n+1} |\alpha_i| \left|R_{\mathbf{k}}\left(y_i\left(\mathbf{k+1}\right)\right)\right|$, we get

$$\eta\left(\mathbf{k}\right) := \left|E^{(\mathbf{k+1})}\right| - \left|E^{(\mathbf{k})}\right| = \sum_{i=0}^{n+1} |\alpha_i| \left|R_{\mathbf{k}}\left(y_i\left(\mathbf{k+1}\right)\right)\right| - \left|E^{(\mathbf{k})}\right|$$

and, from the second formula in [2.10]

$$\sum_{i=0}^{n+1} |\alpha_i| = 1;$$

hence, it holds that

$$\sum_{i=0}^{n+1} |\alpha_i| \left|E^{(\mathbf{k})}\right| = \left|E^{(\mathbf{k})}\right|,$$

i.e.

$$\begin{aligned}\eta\left(\mathbf{k}\right) &= \left|E^{(\mathbf{k+1})}\right| - \left|E^{(\mathbf{k})}\right| = \sum_{i=0}^{n+1} |\alpha_i| \left|R_{\mathbf{k}}\left(y_i\left(\mathbf{k+1}\right)\right)\right| - \sum_{i=0}^{n+1} |\alpha_i| \left|E^{(\mathbf{k})}\right| \\ &= \sum_{i=0}^{n+1} |\alpha_i| \left(\left|R_{\mathbf{k}}\left(y_i\left(\mathbf{k+1}\right)\right)\right| - \left|E^{(\mathbf{k})}\right|\right).\end{aligned}$$

Among the $y_i^{(\mathbf{k+1})}$'s we can find the absolute maximum or minimum of $R_{\mathbf{k}}$. Assume that for $i = j$ we have an absolute maximum, i.e.

$$y_j\left(\mathbf{k+1}\right) := \arg \max_{x \in [a,b]} R_{\mathbf{k}}\left(x\right)$$

which obviously depends on $\mathbf{k}$. Since, $\sum_{i=0}^{n+1} |\alpha_i| \left(\left|R_{\mathbf{k}}\left(y_i\left(\mathbf{k+1}\right)\right)\right| - \left|E^{(\mathbf{k})}\right|\right)$ is the sum of positive terms, it exceeds any of its summands. Hence,

$$\eta\left(\mathbf{k}\right) = \left|E^{(\mathbf{k+1})}\right| - \left|E^{(\mathbf{k})}\right| = \sum_{i=0}^{n+1} |\alpha_i| \left(\left|R_{\mathbf{k}}\left(y_i\left(\mathbf{k+1}\right)\right)\right| - \left|E^{(\mathbf{k})}\right|\right)$$

$$= |\alpha_0| \left(\left|R_{\mathbf{k}}\left(y_i\left(\mathbf{k+1}\right)\right)\right| - \left|E^{(\mathbf{k})}\right|\right) + ... + |\alpha_{j-1}| \left(\left|R_{\mathbf{k}}\left(y_{j-1}(\mathbf{k+1})\right)\right| - \left|E^{(\mathbf{k})}\right|\right) +$$

$$+ \left(\left\|R_{\mathbf{k}}\right\|_{\infty} - \left|E^{(\mathbf{k})}\right|\right) + |\alpha_{j+1}| \left(\left|R_{\mathbf{k}}\left(y_{j+1}(\mathbf{k+1})\right)\right| - \left|E^{(\mathbf{k})}\right|\right) +$$

$$... + |\alpha_{n+1}| \left(\left|R_{\mathbf{k}}\left(y_{n+1}(\mathbf{k+1})\right)\right| - \left|E^{(\mathbf{k})}\right|\right)$$

$$\geq |\alpha_j| \left(\left\|R_{\mathbf{k}}\right\|_{\infty} - \left|E^{(\mathbf{k})}\right|\right).$$

We isolate the term $\left\|R_{\mathbf{k}}\right\|_{\infty}$.

$$\eta\left(\mathbf{k}\right) \geq |\alpha_j| \left\|R_{\mathbf{k}}\right\|_{\infty} - |\alpha_j| \left|E^{(\mathbf{k})}\right|$$

i.e.

$$\frac{\eta\left(\mathbf{k}\right)}{|\alpha_j|} + \left|E^{(\mathbf{k})}\right| \geq \left\|R_{\mathbf{k}}\right\|_{\infty}.$$

Since $E^* \geq \left|E^{(\mathbf{k})}\right|$,

$$\frac{\eta\left(\mathbf{k}\right)}{|\alpha_j|} + E^* \geq \frac{\eta\left(\mathbf{k}\right)}{|\alpha_j|} + \left|E^{(\mathbf{k})}\right| \geq \left\|R_{\mathbf{k}}\right\|_{\infty}.$$

We have

$$E^* := \left\|R^*\right\|_{\infty} = \max_{x \in [a,b]} \left|f\left(x\right) - P_n^*\left(x\right)\right| \leq \max_{x \in [a,b]} \left|f\left(x\right) - P_n^{(\mathbf{k})}\left(x\right)\right| := \left\|R^{(\mathbf{k})}\right\|_{\infty}.$$

It results in

$$E^* \leq \left\|R_{\mathbf{k}}\right\|_{\infty} \leq \frac{\eta\left(\mathbf{k}\right)}{|\alpha_j|} + E^*.$$

Now for $\mathbf{k} \to \infty$

$$0 \leq |\alpha_j| \text{ and } \eta(\mathbf{k}) \to 0.$$

Since there exists $\varepsilon > 0$ and $\mathbf{k}_0$ such that for all $\mathbf{k} > \mathbf{k}_0$

$$\frac{\eta(\mathbf{k})}{|\alpha_j|} \leq \varepsilon$$

it holds that

$$E^* \leq \|R_{\mathbf{k}}\|_\infty \leq E^* + \varepsilon. \qquad \square$$

THEOREM 2.5.– Let $f \in \mathcal{C}^{(0)}([a,b])$, $\{\mathbf{y}(\mathbf{k}) := (y_1(\mathbf{k}), ..., y_{n+2}(\mathbf{k})) : \mathbf{k} \in \mathbb{N}\}$ be the sequence defined by the algorithm of Remez with $\mathbf{y}(\mathbf{0}) := (y_1(\mathbf{0}), ..., y_{n+2}(\mathbf{0}))$ in $[a,b]$, with

$$a = y_1(\mathbf{0}) < ... < y_{n+2}(\mathbf{0}) = b \text{ and } E^{(\mathbf{0})} \neq 0.$$

Then, $\overline{P}_n^{(\mathbf{k})}$ converges uniformly to P^* and $\overline{P}_n^{(\mathbf{k})}$ has exact degree n for all $\mathbf{k}$.

PROOF.– By contradiction, assume that

$$\left\|P_n^* - P_n^{(\mathbf{k})}\right\|_\infty \nrightarrow 0.$$

Then, there exists some subsequence $\left\{P_n^{(\mathbf{k_j})} : \mathbf{k_j} \in \mathbb{N}\right\}$ in $\left\{P_n^{(\mathbf{k})} : \mathbf{k} \in \mathbb{N}\right\}$, such that

$$\left\|P_n^* - P_n^{(\mathbf{k_j})}\right\|_\infty \geq N > 0.$$

Since, $[a,b]$ is compact and $P_n^{(\mathbf{k})}$ is continuous, for all $\mathbf{k}$, applying the Bolzano–Weierstrass theorem, we may extract from $\left\{P_n^{(\mathbf{k_j})} : \mathbf{k_j} \in \mathbb{N}\right\}$, a subsequence which converges uniformly to some limit T. Hence, there exists $\left\{P_n^{\left(\mathbf{k_{j_r}}\right)} : \mathbf{k_{j_r}} \in \mathbb{N}\right\}$ with

$$\left\|P_n^{\left(\mathbf{k_{j_r}}\right)} - T\right\|_\infty \to 0.$$

Now, $E^* \leq \|R_{\mathbf{k}}\|_\infty \leq E^* + \varepsilon$ and therefore,

$$\|R_{\mathbf{k}}\|_\infty = \left\|f - P_n^{(\mathbf{k})}\right\|_\infty \in [E^*, E^* + \varepsilon]$$

with $\varepsilon > 0$ arbitrary. Now, any subsequence of $\left\{P_n^{(\mathbf{k})} : \mathbf{k} \in \mathbb{N}\right\}$, and so also $\left\{P_n^{(\mathbf{k_{j_r}})} : \mathbf{k_{j_r}} \in \mathbb{N}\right\}$, satisfies

$$\left\|f - P_n^{(\mathbf{k_{j_r}})}\right\|_\infty \in [E^*, E^* + \varepsilon].$$

Hence, as $\mathbf{k_{j_r}} \to \infty$,

$$\|f - T\|_\infty \in [E^*, E^* + \varepsilon], \text{ for all } \varepsilon > 0$$

and therefore

$$\|f - T\|_\infty = E^*.$$

Furthermore, from

$$\left\|P_n^* - P_n^{(\mathbf{k_j})}\right\|_\infty \geq N > 0$$

we have, going to the limit in $\mathbf{k_j} \to \infty$, $P_n^{(\mathbf{k_j})} \to T$ and thus

$$\|P_n^* - T\|_\infty \geq N > 0$$

i.e.

$$T \neq P_n^*.$$

However, both T and P_n^* are best uniform approximations of f. This contradicts uniqueness. □

Finally, it can be shown that the rate of convergence in the Remez algorithm is quadratic with respect to the iteration number **k.** For the proof of the following result, see [DZY 08a].

THEOREM 2.6.– Let $f \in \mathcal{C}^{(2)}([a, b])$. If R^* is standard and is such that

$$\left(\frac{d^2 R^*}{dx^2}\right)_{x = x_i^*} \neq 0$$

on the equioscillation points $x_i^*, i = 0, ..., n+1$, then, there exist some $\rho \neq 0$ such that

$$\frac{\left\|R^{(\mathbf{k+1})}\right\|_\infty - E^*}{\left(\left\|R^{(\mathbf{k})}\right\|_\infty - E^*\right)^2} \leq \rho$$

for $\mathbf{k} \in \mathbb{N} \setminus \{0\}$.

3

Constrained Polynomial Approximation

3.1. Introduction and examples

This chapter considers the case when the approximating functions needs to satisfy a number of numerical constraints. The function f to be approximated is a continuous function defined on a bounded interval $[a, b]$; the approximation is performed in a finite dimensional linear space of continuous functions on $[a, b]$ and uniformity is maintained with respect to the uniform norm of functions. The context is, therefore, similar to that of Chapter 1.

The context of this chapter can be summarized as follows. The function f is to be approximated uniformly on $[a, b]$. We have various types of constraints which force us to measure f only on $n+1$ points x_0 , .., x_n in $[a, b]$. Based on these measurements, we intend to produce a function φ, which is defined on $[a, b]$ and uses the measured values of f. Now, assume that φ is chosen as a polynomial with degree n. So, any constraints of the form $\varphi(z_i) = y_i, 0 \leq i \leq n$ allows for a unique φ. We treat the Lagrange interpolation scheme, which provides a convergent sequence of functions φs, indexed by the degree n and enjoys good convergence for the uniform rate of approximation, when the nodes are given as the Chebyshev ones. The function φ coincides with f on these x_i.

The uniform error due to the Lagrange interpolation polynomial can be put in correspondence with the minimal uniform error when approximating the function f by a generic polynomial with given degree n, say P^* as described in Chapter 1. It will be seen that the uniform error pertaining with the Lagrange interpolation polynomial with arbitrary nodes is bounded by the uniform error pertaining to P^*, up to a multiplicative constant depending on n, a property valid for all n. The uniform error due to P^* converges to 0 (see example 3.1). The multiplicative constant depends on the nodes; as n increases, this constant diverges; however, a good choice of the nodes x_i makes this growth slow enough in order to entail the convergence of the approximation error to zero.

EXAMPLE 3.1.– Let $f : [a,b] \subset \mathbb{R} \to \mathbb{R}$, $f \in \mathcal{C}^{(k)}([a,b])$. Assume that f can be measured on some finite subset $S = \{x_0, ..., x_n\}$, in $[a,b]$. Denote $y_i = f(x_i)$, $i = 0, ..., n$, the corresponding measurements, which we assume to be exact, without measurement error. We intend to approximate f with a polynomial φ with degree n such that $y_i = f(x_i) = \varphi(x_i)$, $i = 0, ..., n$, and $\lim_{n\to\infty} \sup_{x\in[a,b]} |f(x) - \varphi(x)| = 0$. The approximation criterion is, therefore, the maximum of the error on $[a,b]$. The resulting polynomial φ is the Lagrange interpolation polynomial of f. The polynomial φ is generated by the canonical basis $\{1; ...; x^n\}$.

The above example is a special case of the polynomial interpolation problem which is usually stated as follows:

Let V be a linear space with dimension $n \geq 1$ on $\mathbb{R}$. Let $\mathcal{A}^* := \{\varphi_0, ..., \varphi_i, ..., \varphi_n\}$ be a basis of the dual space V^*, and $y_i \in \mathbb{R}, i = 0, ..., n$. For any choice of the real numbers $(y_0, ..., y_n)$, find some v in V such that $\varphi_i(v) = y_i$. In the above example V is the class of all polynomial with degree less or equal n, the basis $\mathcal{A}^*$ of V^* can be defined as $n+1$ evaluation functions defined through $\varphi_i(v) = v(x_i)$, for a given set of points. The interpolation problem results as the following one:

Given a function f in $\mathcal{C}^{(k)}([a,b])$, find v in V such that $\varphi_i(v) = v(x_i) = f(x_i) = y_i$. Note that in this example, the basis $\mathcal{A}^*$ can be adjusted to the measurement points x_i's.

Clearly, the solution for this problem exists and is defined in a unique way when the n linear forms $\varphi_1, ..., \varphi_n$ are independent.

Let $\mathcal{B} := \{v_0, ..., v_n\}$ be a basis of V.

For v in V, $v = \sum_{i=0}^n a_i v_i$ and therefore, for all $j = 0, ..., n$ it holds $y_j = \varphi_j(v) = \sum_{i=0}^n a_i \varphi_j(v_i)$. Hence,

$$\begin{pmatrix} y_0 \\ . \\ y_j \\ . \\ y_n \end{pmatrix} = \begin{pmatrix} \varphi_0(v_0) & . & \varphi_1(v_i) & . & \varphi_0(v_n) \\ . & . & . & . & . \\ \varphi_j(v_0) & . & \varphi_j(v_i) & . & \varphi_j(v_n) \\ . & . & . & . & . \\ \varphi_n(v_0) & . & \varphi_n(v_i) & . & \varphi_n(v_n) \end{pmatrix} \begin{pmatrix} a_0 \\ . \\ a_i \\ . \\ a_n \end{pmatrix}.$$

The matrix

$$G := \begin{pmatrix} \varphi_0(v_{0.}) & . & \varphi_0(v_i) & . & \varphi_0(v_n) \\ . & . & . & . & . \\ \varphi_j(v_0) & . & \varphi_j(v_i) & . & \varphi_j(v_n) \\ . & . & . & . & . \\ \varphi_n(v_0) & . & \varphi_n(v_i) & . & \varphi_n(v_n) \end{pmatrix}, \qquad [3.1]$$

is called the Gram matrix. The Gram matrix is invertible, since the forms $\varphi_j, j = 0, ..., n$ are independent. Also it holds

$$\begin{pmatrix} a_0 \\ . \\ a_i \\ . \\ a_n \end{pmatrix} = G^{-1} \begin{pmatrix} y_0 \\ . \\ y_j \\ . \\ y_n \end{pmatrix},$$

which is the solution of the problem.

A special choice of the basis $\{l_0, ..., l_n\}$ of V produces a Gram matrix equal to the identity.

This basis should satisfy,

$$\varphi_j(l_i) = \delta_{ij} := \begin{cases} 1 \text{ when } i = j \\ 0 \text{ when } i \neq j \end{cases}$$

where $\mathcal{A}^* := \{\varphi_0, ..., \varphi_i, ..., \varphi_n\}$ is a basis of V^*.

Denote

$$e_i := \begin{pmatrix} 0 \\ . \\ 1^i \\ . \\ 0 \end{pmatrix}, i = 0, ..., n,$$

the canonical basis of $\mathbb{R}^n$. The vector l_i in V are the reciprocal image of e_i by the isomorphism

$$\Psi : V \to \mathbb{R}^n, v \mapsto \Psi(v) = \begin{pmatrix} \varphi_0(v) \\ . \\ \varphi_n(v) \end{pmatrix}.$$

Hence,

$$l_i = \Psi^{-1}(e_i), i = 0, ..., n.$$

with this basis we obtain,

$$\begin{pmatrix} a_0 \\ . \\ a_i \\ . \\ a_n \end{pmatrix} = \begin{pmatrix} y_0 \\ . \\ y_j \\ . \\ y_n \end{pmatrix}, \qquad [3.2]$$

and therefore,

$$v = \sum_{i=0}^{n} y_i l_i.$$

According to the space V and to the basis $\mathcal{A}^*$ of the dual space V^*, we obtain various classical interpolation schemes.

EXAMPLE 3.2.– Let P belong to $\mathcal{P}_n$. For any real a, consider the linear forms

$$\varphi_i : \mathcal{P}_n \to \mathbb{R}; P \mapsto \varphi_i(P) = \left(\frac{d^i P(x)}{dx}\right)_{x=a}, i = 0, ..., n.$$

The interpolation problem writes: for any $(y_0, ..., y_n)' \in \mathbb{R}^n$ find some polynomial P in $\mathcal{P}_n$, such that $\varphi_i(P) = y_i = \left(\frac{d^i f(x)}{dx}\right)_{x=a}$ for $i = 0, .., n$, making use of the φ_is defined above (depending on the very choice of a). The corresponding solution P is the classical Taylor polynomial with degree n in the Taylor expansion of f in a neighborhood of a. The Gram matrix for this problem is given by,

$$G = \begin{pmatrix} 1\ a\ .\ .\ . & a^n \\ 0\ 1\ .\ .\ . & na^{n-1} \\ .\ .\ .\ .\ . & . \\ 0\ .\ .\ i!\ . & . \\ .\ .\ .\ .\ . & . \\ 0\ 0\ .\ 0\ . & n! \end{pmatrix}.$$

The solution for this interpolation problem exists and is unique; it is given by the Taylor polynomial $P(x) := \sum_{i=0}^{n} \frac{y_i}{i!}(x-a)^i$.

3.2. Lagrange polynomial interpolation

We now consider the case when $V = \mathcal{P}_n$ with $\mathcal{A}$ the canonical basis of V. Let the points x_i satisfy $x_0 < x_1 < ... < x_n$ and the system of functions φ_i be given by

$$\varphi_i : \mathcal{P}_n \to \mathbb{R}; p \mapsto \varphi_i(P) = P(x_i),$$

Then, G is the Vandermonde matrix

$$G = \begin{pmatrix} 1 & x_0^1 & . & x_0^j & . & x_0^n \\ . & . & . & . & . & . \\ 1 & x_i^1 & . & x_i^j & . & x_i^n \\ . & . & . & . & . & . \\ . & . & . & . & . & . \\ 1 & x_n^1 & . & x_n^j & . & x_n^n \end{pmatrix}.$$

Let f be a continuous function defined on $[a, b]$ and assume that f can be observed without error on the points x_i defining $y_i := f(x_i)$.

It holds that $\det G = \prod_{i=0,\ldots,n;i<j} (x_j - x_i)$.

DEFINITION 3.1.– The polynomial

$$\mathcal{L}_n(f)(x) := \sum_{i=0}^{n} y_i l_i(x)$$

is the *Lagrange interpolation polynomial* of f on the nodes $(x_0, .., x_{g-1})$; it goes through the points $(x_i, y_i), i = 0, .., n$. The x_i are the interpolation nodes. The polynomials

$$l_i(x) := \prod_{j=0,j\neq i}^{n} \frac{x - x_j}{x_i - x_j},$$

are the *elementary Lagrange polynomials.*

For a given set of knots $\{x_i : i = 0, ..., n\}$ the linear operator

$$\mathcal{L}_n : C^{(0)}[a; b] \to \mathbb{R}_n[x]; \ f \mapsto \mathcal{L}_n(f)$$

where $\mathcal{L}_n(f) = \sum_{i=0}^{n} f(x_i) l_i(.)$ is the interpolation Lagrange operator.

This operator is continuous. Indeed, $\left(\mathcal{C}^{(0)}\left[a;b\right], \|.\|_{\infty}\right)$ is a Banach space (i.e. it is a complete linear space). Clearly, $\mathcal{L}_n$ is continuous at zero; indeed for $\delta > 0$, if

$$\|f\|_{\infty} := \max_{x\in[a,b]} |f(x)| \leq \delta$$

then $|f(x_i)| \leq \delta$, for all i, which implies for all x in $[a,b]$,

$$\begin{aligned}|\mathcal{L}_n(f)(x)| &\leq \delta \sum_{i=0}^{n} |l_i(x)| \leq \delta \sum_{i=0}^{n} \max_{i=0,\ldots,n} |l_i(x)| \\ &\leq \delta (n+1) \max_{i=0,\ldots,n} |l_i(x)| = \delta (n+1) \max_{i=0,\ldots,n} \prod_{j=0,j\neq i}^{n} \frac{|x-x_j|}{|x_i-x_j|} \\ &\leq \delta (n+1) \max_{i=0,\ldots,n} \|l_i\|_{\infty}.\end{aligned}$$

which yields,

$$\|\mathcal{L}_n(f)\|_{\infty} \leq \delta (n+1) \max_{i=0,\ldots,n} \|l_i\|_{\infty}.$$

Choose

$$\delta \leq \frac{\epsilon}{(n+1)\max_{i=0,\ldots,n} \|l_i\|_{\infty}},$$

to obtain that for any positive ϵ there exists some δ such that when $\|f-\mathbf{0}\|_{\infty} < \delta$ then $\|\mathcal{L}_n(f) - \mathbf{0}\|_{\infty} < \epsilon$.

Further, since $\mathcal{P}_n$ is a finite dimensional vector space, $\mathcal{L}_n$ is continuous for any norm.

DEFINITION 3.2.– The norm $|||\mathcal{L}_n|||$ of the operator $\mathcal{L}_n$ is the smallest constant c, such that

$$\|\mathcal{L}_n(f)\|_{\infty} \leq c\|f\|_{\infty}$$

for any f in $\mathcal{C}^{(0)}[a;b]$ with $f \neq 0$.

Hence,

$$|||\mathcal{L}_n||| := \sup_{f\neq 0, f\in\mathcal{C}^{(0)}[a;b]} \frac{\|\mathcal{L}_n(f)\|_{\infty}}{\|f\|_{\infty}},$$

and therefore,

$$|||\mathcal{L}_n||| = \sup_{x\in[a,b]} \sum_{i=0}^{n} |l_i(x)|.$$

Indeed,

$$|\mathcal{L}_n(f)(x)| = \left|\sum_{i=0}^{n} f(x_i)\, l_i(x)\right| \le \sum_{i=0}^{n} |f(x_i)|\,|l_i(x)| \le \sum_{i=0}^{n} \left(\sup_{x\in[a,b]} |f(x_i)|\right) |l_i(x)|$$

$$\le \|f\|_\infty \sum_{i=0}^{n} |l_i(x)| \le \left(\max_{x\in[a,b]} \sum_{i=0}^{n} |l_i(x)|\right) \|f\|_\infty .$$

Taking the sup over x on all terms of the above display yields

$$\|\mathcal{L}_n(f)\|_\infty = \sup_{x\in[a,b]} |\mathcal{L}_n(f)(x)| \le \left(\max_{x\in[a,b]} \sum_{i=0}^{n} |l_i(x)|\right) \|f\|_\infty . \qquad [3.3]$$

The term $\max_{x\in[a,b]} \sum_{i=0}^{n} |l_i(x)|$ is defined since the mapping

$$[a,b] \to \mathbb{R};\, x \mapsto \sum_{i=0}^{n} |l_i(x)|$$

is a continuous function on $[a,b]$.

To conclude that $\max_{x\in[a,b]} \sum_{i=0}^{n} |l_i(x)|$ is, indeed, the smallest constant, it is enough to prove that there exists a function, $f \in \mathcal{C}^{(0)}([a,b])$, for which equality holds in [3.3].

Observe that

$$\left|\sum_{i=0}^{n} f(x_i)\, l_i(x)\right| \le \sum_{i=0}^{n} |f(x_i)|\,|l_i(x)|,$$

when all terms $f(x_i)\, l_i(x)$ have same sign. Furthermore,

$$\sup_{x\in[a,b]} |\mathcal{L}_n(f(x))| \le \left(\max_{x\in[a,b]} \sum_{i=0}^{n} |l_i(x)|\right) \left(\sup_{x\in[a,b]} |f(x_i)|\right),$$

when f is such that for any x_i, it holds $|f(x_i)| = \|f(x)\|_\infty$.

A function, f, which satisfies these requirements is

$$f(x) := \begin{cases} 1 \text{ if, } x = x_i \text{ and } l_i(x) \geq 0 \\ r(x), x \in (x_i; x_{i+1}); i = 0; ..., n-1 \\ -1 \text{ if, } x = x_i \text{ and } l_i(x) < 0 \end{cases},$$

where $r(x)$ is the straight line which passes through $x_i; x_{i+1}$ in $(x_i; x_i)$.

We now state the following definition.

DEFINITION 3.3.– The constant

$$|||\mathcal{L}_n||| = c = \max_{x \in [a,b]} \sum_{i=0}^{n} |l_i(x)| \qquad [3.4]$$

is the Lebesgue constant. It depends on the knots $\{x_i : i = 0, ..., n\}$.

3.3. The interpolation error

Clearly when P is substituted by a function f which does not belong to $\mathcal{P}_n$, and when f obeys $f(x_i) = y_i$, $i = 0, ..., n$, then $f(x)$ does not coincide with $\mathcal{L}_n(f)(x)$. We evaluate the corresponding error,

$$\sup_x |f(x) - \mathcal{L}_n(f)(x)|.$$

3.3.1. *A qualitative result*

THEOREM 3.1.– Let f be a function in $\mathcal{C}^{n+1}[a; b]$, and $\mathcal{P}_n$ be the family of all polynomials with degree less or equal n defined on $[a; b]$. Then,

$$\|f - \mathcal{L}_n(f)\|_\infty \leq (1 + |||\mathcal{L}_n|||) \left(\inf_{P \in \mathcal{P}_n} \sup_{x \in [a;b]} |f(x) - P(x)| \right).$$

PROOF.– For any P in $\mathcal{P}_n$, it holds $P(x) = \mathcal{L}_n(P)(x)$, due to the uniqueness of the Lagrange approximating polynomial and the fact that there exists a unique polynomial with degree n defined through $n + 1$ constraints. Therefore,

$$\begin{aligned} &\|f - \mathcal{L}_n(f)\|_\infty \\ &= \|f - P + P - \mathcal{L}_n(f)\|_\infty \\ &\leq \|f - P\|_\infty + \|P - \mathcal{L}_n(f)\|_\infty \\ &= \|f - P\|_\infty + \|\mathcal{L}_n(P) - \mathcal{L}_n(f)\|_\infty \end{aligned}$$

Now,

$$\begin{aligned}\|f-\mathcal{L}_n(f)\|_\infty &\le \|f-P\|_\infty + \left\|\mathcal{L}_n(f)\,(x_i)_{1\le i\le n}\right\|_\infty \|f-P\|_\infty \\ &\le \inf_{P_n\in\mathcal{P}_n} \|f-P\|_\infty \\ &+ |||\mathcal{L}_n||| \inf_{P\in\mathcal{P}_n} \|f-P\|_\infty \\ &= (1+|||\mathcal{L}_n|||)\left(\inf_{P\in\mathcal{P}_n}\|f-P\|_\infty\right).\end{aligned}$$

□

The Lebesgue constant which appears in [3.4] is clearly a divergent term as n increases. The following two propositions shed light on the rate of divergence with respect to the distribution of the knots:

PROPOSITION 3.1.– For $i = 0, ..., n$, let

$$u_i := a + i\frac{b-a}{n},$$

and

$$t_i := \frac{a+b}{2} - \frac{b-a}{2}\cos\left(\frac{2i+1}{2n}\pi\right), \qquad [3.5]$$

be two sequences in $[a;b]$. We have,

$$\frac{2^n}{4n^2} \le |||\mathcal{L}_{n,u}||| \le 2^n,$$

and for some real α

$$\frac{2}{\pi}\ln n \le |||\mathcal{L}_{n,t}||| \le \alpha \ln n$$

where $\mathcal{L}_{n,u}$ is the Lagrange interpolation operator with nodes u_i (respectively $\mathcal{L}_{n,t}$ with nodes t_i).

PROOF.– The proof follows from some calculus; see [BRU 97]. □

The following result also holds:

PROPOSITION 3.2.–

$$|||\mathcal{L}_{n,u}||| \sim \frac{2^{n+1}}{ne\log n}$$

$$|||\mathcal{L}_{n,t}||| \sim \frac{2\ln n}{\pi} \text{ (Bernstein).}$$

PROOF.– See [BRU 97]. □

The nodes defined in [3.5] are the roots of the Chebyshev polynomial of first kind (see Appendix 2):

$$T_n(x) = \cos(n \arccos x).$$

3.3.2. *A quantitative result*

We first recall the following result.

THEOREM 3.2.– Let f belong to $C^{(n+1)}([a,b])$. Assume that f has at least $n+2$ distinct zeros in $[a,b]$. Then, $f^{(n+1)}$ has at least one zero in (a,b).

PROOF.– If f has at least $n+2$ zeros in $[a,b]$, then f' has at least $n+1$ zeros in $[a,b]$. Since f' is derivable function in (a,b), f'' has at least $(n+1)-1=n$ zeros in $[a,b]$. The same argument holds $(f'')' = f^{(3)}$, etc. hence, $f^{(n+1)}$ has at least one zero in $[a,b]$, since $f^{(n)}$ has at least two zeros. □

We now derive the following result:

THEOREM 3.3.– Let f belong to $C^{(n+1)}[a;b]$, and let $\{x_i : i = 0, ..., n\}$ be the interpolation nodes. Then,

$$\|f - \mathcal{L}_n(f)\|_\infty \le \frac{\sup_{x\in[a;b]} \left|f^{(n+1)}(x)\right|}{(n+1)!} \prod_{i=0}^{n} |x - x_{i,n}|.$$

PROOF.– Consider the function,

$$P(t) := \sum_{i=0}^{n} l_i(t) f(t).$$

Define, for any x in $[a,b]$

$$\Psi_x(t) = f(t) - P(t) - \frac{\prod_{i=0}^{n}(t-x_i)}{\prod_{i=0}^{n}(x-x_i)}(f(x) - P(x)).$$

If $t = x$, then

$$\Psi_x(x) = f(x) - P(x) - (f(x) - P(x)) = 0.$$

Also when $t = x_i$, for $i = 0, ..., n$,

$$\begin{aligned}\Psi_x(x_i) &= f(x_i) - P(x_i) - \frac{\prod_{i=0}^{n}(x_i - x_i)}{\prod_{i=0}^{n}(x - x_i)}(f(x_i) - P(x_i)) \\ &= f(x_i) - \sum_{i=0}^{n} l_i(x_i)f(x_i) = 0\end{aligned}$$

which proves that Ψ_x has at least $n+2$ zeros $t = x, x_0, ..., x_n$. Since $f \in \mathcal{C}^{(n+1)}[a;b]$ and $P \in \mathcal{C}^{(n+1)}[a;b]$, it follows that $\Psi_x \in C^{n+1}([a,b])$. By the above theorem (theorem 3.2), $\Psi_x^{(n+1)}$ has at least one zero in (a,b); therefore, there exists $\widetilde{x} \in (a,b)$ such that $\Psi_x^{(n+1)}(\widetilde{x}) = 0$. Write

$$\begin{aligned}\Psi_x^{(n+1)}(t) &= \left(f(t) - P(t) - \frac{\prod_{i=0}^{n}(t - x_i)}{\prod_{i=0}^{n}(x - x_i)}(f(x) - P(x)) \right)^{(n+1)} \\ &= f^{(n+1)}(t) - P^{(n+1)}(t) - \frac{\left(\prod_{i=0}^{n}(t - x_i)\right)^{(n+1)}}{\prod_{i=0}^{n}(x - x_i)}(f(x) - P(x)).\end{aligned}$$

It holds that $P^{(n+1)}(t) = 0$, since P has degree n, and

$$\left(\prod_{i=0}^{n}(t - x_i)\right)^{(n+1)} = (n+1)!$$

Taking $t = \widetilde{x}$ we obtain

$$f^{(n+1)}(\widetilde{x}) - \frac{(n+1)!}{\prod_{i=0}^{n}(t - x_i)}(f(t) - P(t)) = 0;$$

hence

$$f^{(n+1)}(\widetilde{x})\frac{\prod_{i=0}^{n}(t-x_i)}{(n+1)!} = (f(t) - P(t)) =: R(t) \qquad [3.6]$$

An upper bound, independent upon $\widetilde{x}$, is

$$\|R\|_{\infty} = \sup_{t\in[a,b]} |R(t)| = \frac{\prod_{i=0}^{n}|t-x_i|}{(n+1)!} \sup_{t\in[a,b]} \left|f^{(n+1)}(t)\right|$$

which concludes the proof. □

This latest result shows the way the knots contribute to the interpolation error. Indeed it is made of two terms: 1) the first term depends only upon the regularity of the function f; 2) a term depending on the way the interpolation is handled through $\prod_{i=0}^{n}|x-x_i|$.

3.4. The role of the nodes and the minimization of the interpolation error

Without loss of generality and for sake of notational clearness we assume in this section that $[a,b] = [-1,1]$.

We can see that the function

$$\omega(x_1, ..., x_n) := \prod_{i=0}^{n}|x - x_i|$$

takes its minimal value when

$$x_i = \cos\left(\frac{2i+1}{2n}\pi\right). \qquad [3.7]$$

THEOREM 3.4.– Let $Q(x) = a_0 + a_1x + a_2x^2 + ... + a_{n-1}x^{n-1} + x^n$, be a polynomial with degree n, with coefficient of x^n equal 1. Let $x \in [-1;1]$. It then holds that

$$\sup_{x\in[-1;1]}\left(\prod_{i=0}^{n}\left|x - \cos\left(\frac{2i+1}{2n}\pi\right)\right|\right) \leq \sup_{x\in[-1;1]}|Q(x)|.$$

PROOF.– Let $\widetilde{T}_n := \frac{1}{2^n} T_n$, where T_n is the Chebyshev polynomial with degree n. The roots of $\widetilde{T}_n$ coincide with the ones of T_n; they are defined through $t_i := \cos\left(\frac{2i+1}{2n}\pi\right)$, for $1 \leq i \leq n-1$. We, thus, have

$$\widetilde{T}_n(x) = (x - t_0)...(x - t_{n-1}),$$

and

$$\sup_{x\in[-1,1]} \left|\widetilde{T}_n(x)\right| = \frac{1}{2^n} \sup_{x\in[-1;1]} |T_n(x)| = \frac{1}{2^n}, \text{ since } \sup_{x\in[-1;1]} |T_n(x)| = 1.$$

Evaluate $\left|\widetilde{T}_n(x)\right|$ on the x, where $\left|\widetilde{T}_n\right|$ reaches its maximal local values (in fact all those values are equal to $1/2^n$), namely $\widetilde{x}_k = \cos\left(\frac{k\pi}{n}\right)$, $k = 0, ..., n$.

To complete the proof, we show that

$$\frac{1}{2^n} \leq \sup_{x\in[-1,1]} |Q(x)|.$$

By contradiction, assume that

$$\frac{1}{2^n} > \sup_{x\in[-1,1]} |Q(x)|, \text{ i.e. } \frac{1}{2^n} > |Q(x)|, \text{ for all } x,$$

which is to say

$$-\frac{1}{2^n} < Q(x) < \frac{1}{2^n}, \text{ for all } x.$$

Since Q and $\widetilde{T}$ share the same degree, it follows that the polynomial

$$Q(x) - \widetilde{T}_n(x) = \left(s^n + \sum_{i=0}^{n-1} a_i x^i\right) - \left(x^n + \sum_{k=0}^{n-1} b_k x^k\right) = \sum_{i=0}^{n-1} a_i x^i - \sum_{k=0}^{n-1} b_k x^k$$

has degree $n-1$.

Evaluate the polynomial $Q - \widetilde{T}_n$ on the $\widetilde{x}_k$.

$$(Q - \widetilde{T}_n)(\widetilde{x}_k) = Q(\widetilde{x}_k) - \frac{(-1)^k}{2^n} = \begin{cases} Q(\widetilde{x}_k) - \frac{1}{2^n} \text{if } k \text{ is even} \\ Q(\widetilde{x}_k) + \frac{1}{2^n} \text{if } k \text{ is odd} \end{cases}$$

We assumed $-\frac{1}{2^n} < Q(x) < \frac{1}{2^n}$, for all x.

For even values of k taking $\frac{1}{2^n}$ as an upper bound for $Q(x)$, we obtain

$$(Q-\widetilde{T})(\widetilde{x}_k) < \frac{1}{2^n} - \frac{1}{2^n} = 0.$$

Similarly, for odd values of k and using the lower bound $-1/2$ it holds

$$(Q-\widetilde{T}_n)(\widetilde{x}_k) > -\frac{1}{2^n} + \frac{1}{2^n} = 0.$$

Hence,

$$(Q-\widetilde{T}_n)(\widetilde{x}_k)(Q-\widetilde{T}_n)(\widetilde{x}_{k+1}) < 0, \text{for all } k \in 0, ..., n-1.$$

This implies that $Q-\widetilde{T}$ changes sign at each $\widetilde{x}_k$, $k = 0, ..., n$. Therefore, it has $n+1$ roots, a contradiction since its degree is $n-1$. Hence,

$$\sup_{x\in[-1;1]} |T(x)| \leq \sup_{x\in[-1;1]} |Q(x)|, \text{for all } Q, \text{and all } x \text{ in } [-1,1]. \qquad \square$$

3.5. Convergence of the interpolation approximation

Theorem 3.1 shows that the approximation of f relies on the Lebesgue constant (which, in turn, depends on the nodes), and also upon the regularity of f.

Since f is a continuous function defined on a compact interval, there exists a sequence of polynomials, say P_n whose uniform limit is precisely f. It, thus, holds

$$\inf_{P_n\in\mathcal{P}_n} \|f-P_n\|_\infty \to 0$$

as n tends to infinity.

However, this does not imply that

$$\|f-\mathcal{L}_n(f)\|_\infty \to 0.$$

We now explore sufficient conditions pertaining on the nodes and on the regularity of f which makes

$$(1+|||\mathcal{L}_n|||)\left(\inf_{P_n\in\mathcal{P}_n} \|f-P_n\|_\infty\right)$$

a well-defined quantity as n tends to infinity.

Clearly, to make $(1+|||\mathcal{L}_n|||)\left(\inf_{P_n\in\mathcal{P}_n}\|f-P_n\|_\infty\right)=\infty\times 0$ converge to zero, we need to make $1+|||\mathcal{L}_n|||$ go to infinity slowly so that $\inf_{P_n\in\mathcal{P}_n}\|f-P_n\|_\infty$ can go to 0 fast enough, depending on the choice of P_n and of the regularity of f. A natural choice for the nodes is the Chebyshev system of points, defined in formula [3.7]. They make $|||\mathcal{L}_n|||$ grow with the slowest rate. This is established in the two following theorems:

THEOREM 3.5 (Lebesgue).– Let $\omega_f(\delta)$ denote the uniform modulus of continuity of f. With $\{x_{i,n}: i=0,...,n\}$ the Chebyshev nodes in $[a,b]$, whenever $\lim_{\delta\to\infty}\omega_f(\delta)\ln\delta=0$, then $\|f-\mathcal{L}_n(f)\|_\infty\to 0$.

PROOF.– See [ERD 58b]. □

THEOREM 3.6 (Bernstein).– When f belongs $\mathcal{C}^{(1)}([a,b])$,and $\{x_{i,n}: i=0,...,n\}$ are the Chebyshev nodes in $[a,b]$, $\|f-\mathcal{L}_n(f)\|_\infty\to 0$.

PROOF.– See [ERD 43] □

The results due to Faber prove that the inadequate choices for the nodes produce divergence of the uniform error committed substituting f by its Lagrange approximation.

Indeed, it holds (see [RIV 74, p. 13] or [FAB 14]).

THEOREM 3.7.– Let $\{x_i; i\geq 0\}$ be a sequence of points in $[a,b]$. Then, the Lagrange interpolation operator $\mathcal{L}_n$ built upon the x_is, $0\leq i\leq n$, satisfies, for all $n\geq 1$

$$|||\mathcal{L}_n|||\geq\frac{\log n}{2(\pi+2)}.$$

Furthermore, there exists a function f in $\mathcal{C}^{(0)}([a,b])$ such that $\mathcal{L}_n(f)$ does not converge uniformly to f on $[a,b]$.

It remains to explain why the Chebyshev nodes $\{t_i\}_{i=0,..,n;n\in\mathbb{N}}$ lead a convergent scheme while the uniform nodes $\{u_i\}_{i=0,..,n;n\in\mathbb{N}}$ produce a divergent scheme. The answer to this question is dervied from Runge, in a theorem from 1902. To clarify this point, we now examine the so-called Runge phenomenon.

3.6. Runge phenomenon and lack of convergence

The Runge phenomenon can be captured through the following example:

Let

$$f:[-1;1]\to\mathbb{R}; x\mapsto f(x)=\frac{1}{1+25x^2} \quad [3.8]$$

which is a continuous function on its domain.

It also holds, and this proves that the continuity of f is not sufficient for the convergence to 0 of the supremum error, that

$$\sup_{x\in[-1;1]}\left|\frac{1}{1+25x^2}-\sum_{i=0}^{n}\frac{1}{1+25\left(-1+\frac{2k}{n}\right)^2}l_i\left(-1+\frac{2k}{n}\right)\right|\nrightarrow 0$$

meaning that the resulting maximal error in the approximation of f through the Lagrange interpolation polynomial at knots $u_{i,n} := -1+\frac{2k}{n}$ in $[-1;1]$ does not converge to 0. The reason for this fact is clear when we embed the domain of f in the complex plane. Consider the function $\widetilde{f}$ defined through,

$$\widetilde{f}:\mathbb{C}\to\mathbb{C}; z\mapsto f(z)=\frac{1}{1+25z^2}.$$

The restriction of $\widetilde{f}$ to the real axis coincides with f. It has two poles, one on $z=-i/5$, while the other one on $z=i/5$ ($i:=\sqrt{-1}$). By its definition, on its nodes, the Lagrange polynomial coincides with the function f.

We first consider some properties of the interpolation polynomial.

Let

$$\pi_n(x) := \prod_{j=0}^{n}(x-x_j).$$

It can be seen, applying the Cauchy theorem, that the Lagrange polynomial $\mathcal{L}_n\left(\widetilde{f}\right)$ can be written as

$$\mathcal{L}_n\left(\widetilde{f}\right)(x)=\frac{1}{2\pi i}\int_\gamma\frac{\widetilde{f}(z)}{z-x}\frac{\pi_n(z)-\pi_n(x)}{\pi_n(z)}dz,$$

for any x in $[-1,1]$, where γ is any curve which excludes the singularities of $\widetilde{f}$ and includes the segment $[-1,1]$. It suffices to observe that

$$\frac{\pi_n(z)-\pi_n(x)}{z-x}$$

is a polynomial in z of degree n and

$$\pi_n(x_k)=(x_k-x_0)\ldots(x_k-x_k)\ldots(x_k-x_n)=0,$$

yielding

$$\frac{1}{2\pi i}\int_{\gamma}\frac{\widetilde{f}(z)}{z-x_k}\frac{\pi_n(z)-\pi_n(x_k)}{\pi_n(z)}dz = \frac{1}{2\pi i}\int_{\gamma}\frac{\widetilde{f}(z)}{(z-x_k)}\frac{\pi_n(z)}{\pi_n(z)}dz$$
$$= \frac{1}{2\pi i}\int_{\gamma}\frac{\widetilde{f}(z)}{z-x_k}dz = \widetilde{f}(x_k),$$

which entails that

$$\frac{1}{2\pi i}\int_{\gamma}\frac{\widetilde{f}(z)}{z-x}\frac{\pi_n(z)-\pi_n(x)}{\pi_n(z)}dz$$

is a polynomial in x which coincides with $\widetilde{f}(x)$ on the knots. This identifies $\mathcal{L}_n\left(\widetilde{f}\right)$.

The interpolation error is then

$$f(x)-\mathcal{L}_n(f)(x) = \frac{1}{2\pi i}\int_{\gamma}\frac{\widetilde{f}(z)}{z-x}dz - \frac{1}{2\pi i}\int_{\gamma}\frac{\widetilde{f}(z)}{z-x}\frac{\pi_n(z)-\pi_n(x)}{\pi_n(z)}dz$$
$$= \frac{1}{2\pi i}\int_{\gamma}\frac{\widetilde{f}(z)}{z-x}\frac{\pi_n(x)}{\pi_n(z)}dz.$$

It follows that the interpolation error depends upon the ratio

$$\frac{\pi_n(x)}{\pi_n(z)}.$$

This sheds light on the role of the nodes in the error. In order that, as $n\to\infty$, $\mathcal{L}_n(f)(x)$ converges to $f(x)$ it is enough to make this ratio go to 0 for any z in γ. For this sake, it is enough that there exists a path γ and a constant $h<1$, such that for any z in γ,

$$\left|\frac{\pi_n(x)}{\pi_n(z)}\right| \le h^n, \text{ for all } n \text{ large.}$$

In this case, as $n\to\infty$,

$$|f(x)-\mathcal{L}_n(f)(x)| = \left|\frac{1}{2\pi i}\int_{\gamma}\frac{\widetilde{f}(z)}{z-x}\frac{\pi_n(x)}{\pi_n(z)}dz\right|$$
$$\le \frac{1}{2\pi i}\int_{\gamma}\left|\frac{\widetilde{f}(z)}{z-x}\frac{\pi_n(x)}{\pi_n(z)}\right|dz \le \frac{h^n}{2\pi i}\int_{\gamma}\left|\frac{\widetilde{f}(z)}{z-x}\right|dz \to 0.$$

Consider the uniform nodes,

$$x_{j,n} = -1 + 2\frac{j}{n}.$$

We will prove that the sequence $\mathcal{L}_n(f)(x)$ converges uniformly to $f(x) = 1/\left(1+25x^2\right)$ only on some interval $[-a,a]$, which is strictly included in $[-1,1]$.

It holds, with those nodes

$$\lim_{n\to\infty} \frac{1}{n}\log|\pi_n(z)| = \frac{1}{2}\mathcal{R}\left(\int_{-1}^{1}\log(z-t)\,dt\right),$$

where $\mathcal{R}(z)$ is the real part of z, which yields

$$\begin{aligned} G(z) &:= \lim_{n\to\infty} \sqrt[n]{|\pi_n(z)|} \\ &= \exp\left\{\frac{1}{2}\mathcal{R}\left((z+1)\log(z+1)-(z-1)\log(z-1)\right)-1\right\}, \end{aligned}$$

Take $z = x$, which entails

$$G(x) := \lim_{n\to\infty} \sqrt[n]{\pi_n(x)} = \frac{1}{e}\sqrt{\frac{(x+1)^{(x+1)}}{(x-1)^{(x-1)}}}. \qquad [3.9]$$

These expressions will provide the asymptotic behavior of $\frac{\pi_n(x)}{\pi_n(z)}$.

We can now state the following result:

THEOREM 3.8 (Runge).– Let a denote some point in $(-1,1)$ such that f has no singularity in

$$\{z\in\mathbb{C} : G(z)\le G(a)\}. \qquad [3.10]$$

Consider the uniformly distributed points$\left\{x_j = -1+2\frac{j}{n} : j = 0,...,n; n\in\mathbb{N}\right\}$. We then have $\mathcal{L}_n(f)(x)\to f(x)$, uniformly on $(-a;a)$ as $n\to\infty$.

PROOF.– Let γ be a path containing both $[-1;1]$ and γ is such that the set

$$\{z \in \mathbb{C} : G(z) \leq G(a)\}$$

lies inside γ and a be such that all singularities of $\widetilde{f}$ are outside γ. The above set is a closed convex set in $\mathbb{C}$ which is bounded by the level set $G(z) = G(a)$. For any x in $(-a; a)$, it holds

$$G(x) < G(a) \leq \min_{z\in\gamma} G(z), \qquad [3.11]$$

where

$$G(x) = \frac{1}{e}\sqrt{\frac{(x+1)^{(x+1)}}{(x-1)^{(x-1)}}}.$$

The strict inequality in [3.11] follows from a direct study of the function G.

But

$$\sqrt[n]{\pi_n(x)} \to \frac{1}{e}\sqrt{\frac{(x+1)^{(x+1)}}{(x-1)^{(x-1)}}} \text{ and } \min_{z\in\gamma} \sqrt[n]{|\pi_n(z)|}) \to_{n\to\infty} \min_{z\in\gamma} G(z).$$

Then for large values of n, using [3.11] for any z in γ

$$\lim_{n\to\infty} \frac{\sqrt[n]{\pi_n(x)}}{\sqrt[n]{\pi_n(z)}} \leq \lim_{n\to\infty} \frac{\sqrt[n]{\pi_n(x)}}{\min_{z\in\gamma} \sqrt[n]{|\pi_n(z)|})} \leq \frac{\frac{1}{e}\sqrt{\frac{(x+1)^{(x+1)}}{(x-1)^{(x-1)}}}}{\min_{z\in\gamma} G(z)} < 1\,. \qquad [3.12]$$

□

The value of a is obtained as follows:

Solve $G(a) = G(i/5)$; then uniform convergence of $f(x) - \mathcal{L}_n(f)(x)$ to 0 holds on $(-a, a)$. An explicit calculation yields $a = 0.7226$.

For the Chebyshev nodes, it holds

$$\lim_{n\to\infty} \sqrt[n]{\pi_n(z)} = \frac{1}{2}\max\left\{\left|z + \sqrt{z^2-1}\right|, \left|z - \sqrt{z^2-1}\right|\right\} \qquad [3.13]$$

from which it can be proved that the uniform convergence of the interpolating Lagrange polynomials to f holds on $[-1, 1]$. We refer to Isaacson and Keller [ISA 94].

3.7. Uniform approximation for $C^{(\infty)}([a,b])$ functions

The Lagrange interpolation scheme can be used for the approximation of the derivatives of a given function. First notice that for any integer j, the function $\mathcal{L}_n\left(f^{(j)}\right)(x)$ coincides with $f^{(j)}$ when f is a polynomial. The following results state that the derivatives of the Lagrange interpolation polynomial of some differentiable function f are good candidates for the approximation of the corresponding derivative of f.

THEOREM 3.9.– When $f \in \mathcal{C}^{\infty}([a,b])$, for any integers h and r,

$$\lim_{n\to\infty} n^r \max_{i=0,1,\ldots,n} \left\| f^{(i)} - \mathcal{L}_n^{(i)}(f) \right\|_{\infty} = 0.$$

PROOF.– See Coatmèlec [COA 66]. □

We state a qualitative result which describes the error of the interpolation approximation scheme with respect to the derivatives of the function.

THEOREM 3.10 (Coatmèlec).– Given the points

$$x_{i,n} = \frac{a+b}{2} - \frac{b-a}{2}\cos\left(\frac{2i-1}{2n+2}\pi\right), i = 1, .., n+1,$$

in $[a,b]$, and given a function f in $\mathcal{C}^{(m)}([a,b])$, assuming that $n \geq 2m-3$, it holds

$$\sup_{x\in[a,b]} \left|f - \mathcal{L}_n(f(x))\right| \leq \left(9 + \frac{4}{\pi}\ln(n+1)\right) \sup_{x\in[a,b]} \left|f^{(m)}(x)\right| \left(\frac{\pi(b-a)}{2(1+n)}\right)^m.$$

PROOF.– See Coatmèlec [COA 66]. □

REMARK 3.1.– When $f \in \mathcal{C}^{\infty}([a,b])$, then for any j and m with $n \geq 2m-3$,

$$\left\| f^{(j)} - \mathcal{L}_n^{(j)}(f) \right\|_{\infty} \leq \left(9 + \frac{4}{\pi}\ln(n+1)\right) \sup_{x\in[a,b]} \left|f^{(m+j)}(x)\right| \left(\frac{\pi(b-a)}{2(1+n)}\right)^m.$$

We now turn to another phenomenon which impacts on the convergence of the Lagrange interpolation scheme, namely the numerical instability of the procedure.

Until now, we have assumed that the function f is known without any error on the nodes. With such a hypothesis, the convergence of the interpolating scheme depends on the regularity of f and on the choice of the nodes. We also mentioned that convergence may not hold even for very regular functions f, for example in $\mathcal{C}^{(\infty)}([-1,1])$, due to the Runge phenomenon (depending, therefore, upon the choice of the nodes).

Lack of convergence may also happen due to some uncertainty on the exact knowledge of the values of f on the interpolation nodes. This topic will be discussed in the next section.

3.8. Numerical instability

For sake of notational clarity, we start our discussion assuming that f is a continuous function on $[0,1]$. The basis is $\mathcal{A} := \{1, x, .., x^n\}$, the canonical basis of the linear space $\mathcal{P}_n$ of all polynomials with degree less or equal n on $[0,1]$. We evaluate the Lagrange interpolation polynomial using $\mathcal{A}$. Let $\{x_0, .., x_n\}$ denote the interpolating nodes in $[0,1]$. The Gram matrix [3.1] takes the form of the Vandermonde matrix

$$G := \begin{pmatrix} 1 & ... & x_0^j & ... & x_0^n \\ ... & ... & ... & ... & ... \\ 1 & ... & x_i^j & ... & x_i^n \\ ... & ... & ... & ... & ... \\ 1 & ... & x_n^j & ... & x_n^n \end{pmatrix}$$

As j increases, all terms x_i^j take similar very small values (all the x_i belong to $(0,1)$); typically when $n \geq 20$, x^j cannot be separated from x^{j+1} in floating arithmetics for $0 < x < 1$. This introduces the first source of strong instability, independently of the lack of knowledge of f on the nodes, depending only on the choice of the Chebyshev system. Good choices for $\mathcal{A}$ exist, avoiding somehow the present source of instability. The choice of the basis $\mathcal{A}$ of $\mathcal{P}_n$ is important in this regard. The elementary Lagrange polynomials suffer of strong instability when the gaps between the nodes is small. This leads to more robust choices for the basis. A commonly used basis of $\mathcal{P}_n$ consists of the Newton polynomials, defined by

$$\begin{aligned} \{\omega_0(x) &= 1, \omega_1(x) = x - x_0, ..., \omega_i(x) \\ &= \prod_{j=0}^{i-1}(x - x_j), ..., \omega_n(x) = \prod_{j=0}^{n-1}(x - x_j)\}, \end{aligned} \qquad [3.14]$$

where

$$x_0 < .... < x_{n-1}$$

are pairwise distinct real numbers. This defines a basis of $\mathcal{P}_n$, called the Newton basis; see [ISA 94].

When the nodes are very close to one another, the rows tend to become linearly dependent, at least numerically.

The main problem related with the instability of the interpolating scheme is when the values of f are known up to some uncertainty on the nodes. This case pertains to all the experimental fields. In this case, instability is due to the fact that the interpolation algorithm results as an ill-conditioned linear system. This can be stated as follows:

$$\left\| \mathcal{L}_n (f)(x) - \mathcal{L}_n \left(\widetilde{f}\right)(x) \right\|_\infty \leq K_n \sup_{i=0,\dots,n} \left(\left| f(x_i) - \widetilde{f}(x_i) \right| \right)$$

where $K_n > 1$ is called the conditionement constant with $\lim_{n\to\infty} K_n = \infty$.

We now provide an explicit expression for K_n.

Let $\{x_0, .., x_n\}$ denote the interpolating nodes in $[a, b]$.

Let $f(x)$ denote the true value of the function and $\widetilde{f}(x)$ an erroneous value due for example to a error resulting from measurement or truncating. Let

$$\varepsilon_i := \widetilde{f}(x_i) - f(x_i), 0 \leq i \leq n$$

be the resulting values of the error on the nodes.

Denote

$$\|\varepsilon\|_\infty := \max_{k=0,\dots,n} |\varepsilon_k|,$$

the maximal error on the nodes (which will be called input errors). We have

$$\sup_{x\in[a,b]} \left| \mathcal{L}_n (f)(x) - \mathcal{L}_n \left(\widetilde{f}\right)(x) \right| = \sup_{x\in[a,b]} \left| \sum_{i=0}^n f(x_i) l_i(x) - \sum_{i=0}^n \widetilde{f}(x_i) l_i(x) \right|$$

$$\leq \sup_{x\in[a,b]} \sum_{i=0}^n \left| f(x_i) - \widetilde{f}(x_i) \right| |l_i(x)|$$

$$\leq \sup_{x\in[a,b]} \sum_{i=0}^n \left(\sup_{i=0,\dots,n} \left(\left| f(x_i) - \widetilde{f}(x_i) \right| \right) \right) |l_i(x)|$$

$$\leq \|\varepsilon\|_\infty |||\mathcal{L}_n|||$$

where $K_n = |||\mathcal{L}_n|||$.

Hence, if $\|\varepsilon\|_\infty \neq 0$ (even when it is very small) the error may propagate and even diverge, since $|||\mathcal{L}_n|||$ is sum of positive terms. A counterexample is as follows:

EXAMPLE 3.3.– Let

$$f(x) := \sin(2\pi x) \text{ with } x \text{ in } [-1,1].$$

Choose 22 equispaced nodes. Modify $f(x)$ randomly on the nodes such that

$$\max_{i=0,..,21} \left| f(x_i) - \widetilde{f}(x_i) \right| \sim 9.5 \times 10^{-4}$$

where $\widetilde{f}$ denotes the modified function. Then,

$$\max_{x\in[-1,1]} \left| \mathcal{L}_n(f)(x) - \mathcal{L}_n\left(\widetilde{f}\right)(x) \right| \sim 3.134$$

where the gap is particularly severe at the end points of the interval; see [QUA 07].

Therefore, the instability ruins the convergence of the Lagrange interpolation for any system of knots and any function f, even when very regular.

REMARK 3.2.– Only in the linear case does the numerical stability of the Lagrange polynomial hold. Consider two knots, x_0 and x_1; the resulting Lagrange polynomial is

$$\sum_{i=0}^{1} f(x_i)\, l_i(x) = \frac{x - x_1}{x_0 - x_1} f(x_0) + \frac{x - x_0}{x_1 - x_0} f(x_1)$$

where

$$\sum_{k=0}^{1} |l_i(x)| = \left| \frac{x - x_1}{x_0 - x_1} \right| + \left| \frac{x - x_0}{x_1 - x_0} \right|.$$

For $x_0 < x < x_1$, we find that $|||\mathcal{L}_n||| < 1$, implying stability.

From the above calculus it appears that the error depends on the nodes. The order of magnitude of the optimal lower bound for the Lebesgue constant is provided by the following result.

THEOREM 3.11 (Erdös).– Let $\mathcal{S} := \{(x_1, ..., x_n) \in \mathbb{R}^n : x_{(1)} < ... < x_{(n)}, n \in \mathbb{N}\}$. It then holds

$$\left(\inf_{(x_1,...,x_n)\in\mathcal{S}} |||\mathcal{L}_n|||\right) \sim \frac{2\ln n}{\pi}.$$

PROOF.– See Erdös [ERD 58a]

This result states that asymptotically the Lebesgue constant associated with the Chebychev nodes is optimal.

REMARK 3.3.– The Chebyshev nodes enjoy the following properties: 1) for any function f it provides a minimal error for the interpolation; 2) the resulting interpolation is asymptotically optimal in term of the stability with respect to input errors; 3) with additional hypotheses on $f \in C^0([a,b])$ convergence holds for the family of interpolating polynomials (see theorem 3.5).

REMARK 3.4.– To avoid the Runge phenomenon and the instability of the interpolation scheme, which depend heavily upon the degree of the interpolating polynomial, it is possible to use splines.

We will now discuss the use of the Lagrange interpolation with respect to the methods of splines, which consists of subdividing the domain of f in a number of small intervals on which f is approximated by a local polynomial with low degree. This reduces the instability phenomenon. The polynomials are fitted in such a way that they produce a smooth approximation also at the boundaries of the intervals of the partition of the support of f. For example, choosing parabolas for the local approximations implies a strong dependency of the approximation scheme with respect to the slope of the approximation on the first boundary. This lack of robustness of the spline scheme can be improved using cubic curves. The regularity conditions at the boundaries amount to state two conditions pertaining to the passage through the nodes, plus two conditions on the second derivative of the cubic functions. The curvature radius r of the function f is given by

$$r := \frac{\sqrt{\left(1 + (f')^2\right)^3}}{f''}$$

which imposes the continuity of the second derivative of the approximating function at the nodes. On the contrary, the Lagrange interpolation scheme is global. In forthcoming chapters, we will use polynomials with low degree Chebyshev nodes, since we will consider smooth functions f, typically real analytic functions. This will make the Runge phenomenon disappear. As will be developed in later chapters, the Lagrange scheme produces a simple way to produce an experimental design with some optimality, both in the interpolation and in the extrapolation.

3.9. Convergence, choice of the distribution of the nodes, Lagrange interpolation and splines

From the standpoint of the convergence properties of the approximating function, it is impossible to determine a distribution of the nodes which enjoys some uniformity property with respect to the convergence. According to the Faber theorem, for any system of nodes, there exists at least a function $f \in \mathcal{C}^{(0)}([a,b])$ for which the interpolation diverges. Henceforth, we cannot provide a universal system of nodes which ensures convergence of the interpolation design for all f in $\mathcal{C}^{(0)}([a,b])$.

The context changes when we move to $\mathcal{C}^{(1)}([a,b])$. The Chebyshev system of nodes provides the unique optimal system. Outside $\mathcal{C}^{(1)}([a,b])$ we ought to consider functions with various types of behaviors and examine the convergence when varying the system of nodes. The resulting comparison of the interpolation errors for the same number of nodes indicates the appropriate distribution of the nodes for a given class of functions.

Consider the case of discontinuous functions on one hand, while the class of spline functions on the other.

In the first case (consider the prototype function $x \to \frac{x}{|x|}$ in $[-5;4]$), the interpolation error committed using the Lagrange polynomials and the Chebyshev nodes decreases as the number of nodes increases, but does not tend to 0. With the uniformly distributed nodes the interpolation error increases with the number of nodes. In this case, the Chebyshev nodes improve on the equidistant ones.

A prototype of functions with rough variations is

$$f : [\varepsilon; 4] \to \mathbb{R}, f(x) := \frac{1}{1 + \ln x}, \varepsilon > 0.$$

In this case, the approximation does not converge; however, also using the Lagrange polynomials, the Chebyshev nodes improve on the equidistant ones; indeed, increasing the number of nodes, with the first system the error stabilizes to a positive value, and with the second system it diverges.

In the case when we consider spline functions, the distribution of the nodes varies in the partition of the domain, changing from one subinterval to the other. By changing the partition, the distribution of the nodes changes completely. However, in the case of spline functions, the approximation depends on the distribution of the nodes only marginally.

From the standpoint of the choice of the approximation, we intend to prove that the splines are to be preferred to the Lagrange polynomials.

In the case of the Runge function, we know that strong oscillations of the Lagrange polynomial at the frontiers of the interval are due to singularities of f in the complex plane, close to the boundaries. These wild oscillations increase together with the number of nodes and can be controlled using the Chebyshev nodes. This choice makes the interpolation error to tend to some other way to find convergence. It does so by using various subsets of the domain of f, and different polynomials with limited fixed degree (usually less than 4). This is the choice of the splines. Convergence is obtained making the length of each subinterval tend to 0, thereby increasing the number of polynomials whose maximal degrees remain fixed.

Clearly, in the case where the oscillation phenomenon does not occur, the Runge phenomenon is not present.

Comparing the use of the Lagrange polynomials evaluated on the Chebyshev nodes with the linear splines evaluated on equidistant nodes on each of the subintervals the following facts can be observed:

1) in both cases the interpolation error goes to 0;

2) for the Lagrange polynomials the error is maximal at the frontier of the domain;

3) it is maximum at the center of the domain for the splines.

The function has rough variations at the center of the domain, and the linear splines cannot compete there efficiently. Clearly, this error reduces when linear splines are substituted by cubic splines.

When considering splines with low degrees it is clear that a function with changes of curvature at various points of its domain will be approximated with better accuracy using Lagrange polynomials. In the case of simple trigonometric functions, it can be observed that the Lagrange approximation provides a better fit than the splines, with convergence of the error to zero, when built on the Chebychev nodes.

Analyzing now the convergence of the interpolation error and comparing the Lagrange approximation on the Chebychev nodes with the linear and cubic splines on the equidistant nodes, we may conclude that there is no universal answer to the best approximating scheme.

From a series of numerical simulations we may see that in some cases the error in the first scheme goes to zero more rapidly than with the second one. However, for discontinuous functions such as $x \to x/|x|$, the Lagrange polynomials do not converge and the splines produce an error which seems to converge, although slowly.

PART 2

Optimal Designs for Polynomial Models

4

Interpolation and Extrapolation Designs for the Polynomial Regression

4.1. Definition of the model and of the estimators

We assume that we are given an interval where the explanatory variable x takes its value; for simplicity assume that x belongs to $[-1, 1]$. In this interval the response Y can be observed. An additive noise causes Y to be only partly related to the input x. This noise is assumed to be independent of the value of x, which is commonly referred to as a homoscedastic hypothesis on the model. For a given input x the measure $Y(x)$ can be written as,

$$Y(x) = P(x) + \varepsilon$$

where P is some unknown function and the generic real valued random variable ε has some unknown distribution; however, it is assumed that the first two moments of ε are finite. The function P might be defined on a larger interval than $[-1, 1]$. All possible measurements of P can only be achieved on $[-1, 1]$. It may occur that we are interested in some estimation of $P(x)$, for some x, where P is not measured; when x belongs to $[-1, 1]$ this is an interpolation problem. At times, we may be interested in some approximation of $P(x)$ for x outside $[-1, 1]$; this is an extrapolation problem.

We will now discuss optimal designs in those contexts. Defining a design results in a two fold description. Firstly, it is based on a set of measurement points, say $x_0, .., x_{g-1}$ in $[-1, 1]$. Those are the nodes of the design. Secondly, for any node x_j, we define an integer n_j, which is the number of replicates of the measurement performed under the condition x_j. We, thus, inherit the n_j measurements $Y_1(x_j), .., Y_{n_j}(x_j)$. Those measurements are supposed to be independent. Note that we do not assume any probabilistic structure on the x_j which, therefore, will not be considered as sampled under any device. The x_j are determined by the experimenter and their choice will follow from a strictly deterministic procedure.

Obviously, this simple model will allow for a simple estimate of $P(x_j)$ for all j, assuming without loss of generality that the error ε has expectation 0.

The design is, therefore, defined by the family of the nodes (their number g is fixed by the experimenter) and the so-called frequencies which we define now.

Obviously, the total number of experiments is limited, for reasons which have to do with the context of the study. Those reasons might be related to the cost of each individual experiment, or by other considerations. For example, in phase 1 clinical trials it is usually assumed that only very few patients can be eligible for the trial. Call n the number of trials to be performed. The resulting constraint on n_j is therefore,

$$n_1 + .. + n_{g-1} = n.$$

Let us now formally define this model.

For any $i = 1, .., g-1$, x_i is a node and $y_k(x_i)$ is the th measurement of $Y(x_i)$, when k runs in $1, .., n_i$. Therefore,

$$\begin{cases} y_1(x_i) = P(x_i) + \varepsilon_{1,i} \\ \ldots\ldots\ldots\ldots\ldots\ldots\ldots \\ y_{n_i}(x_i) = P(x_i) + \varepsilon_{n_i,i} \end{cases}$$

where $n_i > 1$ and $n_i \in \mathbb{N}$, together with $n := \sum_{i=1}^{g-1} n_i$ where n is fixed and $i = 0, ...n$. Obviously the r.v's $\varepsilon_{j,i}$, $1 \leq j \leq n_i$, $i = 0, ..., g-1$, are not observed. They are independent and identically distributed (i.i.d.) copies of a generic random variable (RV) ε. Furthermore $E(\varepsilon) = 0, var(\varepsilon) = \sigma^2$.

We assume that P is a polynomial with known degree $g - 1$. Therefore, it is completely determined if known the values of P in g distinct points. Note that the knowledge of g is an important and strong assumption. Furthermore, denote

$$I := \{x_0 < ... < x_{g-1}\} \subset [-1, 1]$$

as the family of nodes.

The aim of this chapter is to discuss the operational choice of the design; we will, thus, propose some choices for the nodes and the so-called frequencies n_j/n which, all together, define the design. This will be achieved discussing some concept of optimality.

When no random effect is present, existence and uniqueness of the solution of the linear system with g equations and g variables $\theta := (\theta_0, ..., \theta_{g-1})$,

$$y(x_i) = \sum_{j=0}^{g-1} \theta_j x_i^j, (x_0, ..., x_{g-1}) \in [-1, 1]^g, 0 \le i \le g-1$$

allows to identify the function P at any x in $\mathbb{R}$. Changing the canonical basis in $\mathcal{P}_{g-1}$ into the family of the elementary Lagrange polynomials

$$l_i(x) := \prod_{j=0, j \ne i}^{g-1} \frac{x - x_j}{x_i - x_j} \qquad [4.1]$$

yields

$$P(x) = \sum_{i=0}^{g-1} l_i(x) P(x_i).$$

In the present random setting, $P(x_i)$ is unknown. This suggests to consider the estimator $\widehat{P(x)}$ of $P(x)$ defined by

$$\mathcal{L}_n(\widehat{P})(x) := \sum_{i=0}^{g-1} \widehat{P(x_i)}\, l_i(x),$$

where $\widehat{P(x_i)}$, denotes some estimate of P on a generic node x_i.

Turning to the estimate of $P(x_i)$ namely the simplest one, defined by

$$\widehat{P(x_i)} := \overline{Y}(x_i) := \frac{1}{n_i} \sum_{j=1}^{n_i} Y_j(x_i),$$

which solves

$$\widehat{P(x_i)} = \arg\min_{\mu \in \mathbb{R}} \sum_{j=1}^{n_i} (Y_j(x_i) - \mu)^2$$

provides the optimal linear unbiased estimator of $P(x_i)$. It follows that $\widehat{P(x)} := \mathcal{L}_n(\widehat{P})(x)$ is unbiased since for all $x \in \mathbb{R}$,

$$E\left(\mathcal{L}_n(\widehat{P})(x)\right) := \sum_{i=0}^{g-1} E\left(\widehat{P(x_i)}\right) l_i(x) = \mathcal{L}_n(P)(x) = P(x).$$

Since $\mathcal{L}_n(\widehat{P})(x)$ is linear with respect to the parameters $P(x_i)$, $i = 0, ..., g-1$, using Gauss Markov Theorem, $\mathcal{L}_n(\widehat{P})(x)$ is optimal, i.e. has minimal variance.

The variance of the estimator $\mathcal{L}_n(\widehat{P})(x)$ is

$$var\left(\mathcal{L}_n(\widehat{P})(x)\right) = \sum_{i=0}^{g-1} var\left(\widehat{P(x_i)}\right)(l_i(x))^2 = \sigma^2 \sum_{i=0}^{g-1} \frac{(l_i(x))^2}{n_i}, \quad [4.2]$$

which depends explicitly on the frequency n_i of the observations of P on the nodes x_is.

We now get to the formal definition of a design.

The set

$$\left\{((n_0, ..., n_{g-1}), (x_0, ..., x_{g-1})) \in \mathbb{N}^g \times [-1,1]^g : n = \sum_{i=0}^{g-1} n_i,\ n \text{ fixed}\right\}$$

determines a discrete probability measure ξ with support I, a finite subset in $[-1,1]$, by

$$\xi(x_i) := \frac{n_i}{n},\ i = 0, ..., g-1.$$

Turning to [4.2] we observe that the accuracy of the design depends on the point x where the variance of $\mathcal{L}_n(\widehat{P})(x)$ is calculated.

Since all estimators of the form $\mathcal{L}_n(\widehat{P})(x)$ are unbiased, their accuracy depends only on their variance, which, in turn, depends both on x and on the measure ξ. Optimizing on ξ for a given x turns to an optimal choice for I and for the family of the n_is, under the constraint

$$n_0 + .. + n_{g-1} = n.$$

Such designs ξ_x are called Hoel-Levine extrapolation designs when x lies outside $[-1,1]$. When x belongs to $[-1,1]$ then clearly the optimal design for the criterion of the variance of the estimator of $P(x)$ results in performing all the n measurements at point x. The non-trivial case is when the optimality is defined through a control of the uniform variance of the estimator of $P(x)$, namely when ξ should minimize

$$\sup_{x\in[-1,1]} var_\xi \mathcal{L}_n(\widehat{P})(x).$$

Those designs ξ are called interpolation designs, or Guest or Legendre designs.

The notation to be kept is as follows. The set of all probability measures on the interval $[-1,1]$ supported by g distinct points in the interval $[-1,1]$ is denoted $\mathcal{M}^*_{[-1,1]}$, which, therefore, is the class of all designs.

The purpose of this chapter is to present a unified view on this classical field which links the theory of the uniform approximation of functions and the statistical theory of experimental designs, unifying notation and concepts. Its content comes mostly from the seminal works by Hoel and Levine [HOE 64] and Guest [GUE 58]. Some other important references in this context are those of Kiefer and Wolfowitz [KIE 64] and Studden [STU 68].

4.2. Optimal extrapolation designs: Hoel–Levine or Chebyshev designs

We consider the problem of estimating $P(x)$, for any fixed x in the interval $[c,-1]$ for some $c<-1$. More generally, the optimal design which is obtained in this section is valid for any c such that $|c|>1$.

As seen previously, since $\mathcal{L}_n(\widehat{P})(x)$ is an unbiased estimate of $P(x)$, a natural criterion for optimality in the class of all unbiased linear estimates is the variance.

We, therefore, consider the problem

$$\xi_x^* := \arg\min_{\xi\in\mathcal{M}^*_{[-1,1]}} var\left(\mathcal{L}_n(\widehat{P})(x)\right).$$

Denoting generically $\mathbf{n}^* := \left(n_0^*,...,n_{g-1}^*\right)$ and $\underline{\mathbf{x}}^* := \left(x_0^*,...,x_{g-1}^*\right)$, $\mathbf{n} := (n_0,...,n_{g-1})\in\mathbb{N}^g$ and $\underline{\mathbf{x}} := (x_0,...,x_{g-1})$ this problem can thus be written as

$$\left\{\begin{array}{c}(\mathbf{n}^*,\underline{\mathbf{x}}^*) = \arg\min_{(\mathbf{n},\underline{\mathbf{x}})}\sum_{i=0}^{g-1}\frac{(l_i(x))^2}{n_i}\\ \mathbf{n}:\ \sum_{i=0}^{g-1} n_i = n,\, n \text{ fixed}\\ x<-1\leq x_0<...<x_{g-1}\leq 1, \underline{\mathbf{x}}\in[-1,1]^g\end{array}\right. .$$

This is an integer programming problem (with respect to $\mathbf{n}$) with an inequality constraint in $\mathbb{R}$. As done commonly, we find a proxy to the solution, considering the n_is as real numbers. The resulting solution $\left(n_0^*,...,n_{g-1}^*\right)$ will be substituted by the integer part of each of the n_i^*'s.

We, therefore, get to the following optimization problem in the $2g$ real variables

$$\left\{\begin{array}{c}\left(\left(w_0^*,...,w_{g-1}^*\right),\underline{\mathbf{x}}^*\right) = \arg\min_{((w_0,...,w_{g-1}),\underline{\mathbf{x}})}\sum_{i=0}^{g-1}\frac{(l_i(x))^2}{w_i}\\ (w_0,...,w_{g-1})\in\mathbb{R}^g, w>0,\ \sum_{i=0}^{g-1} w_i = n,\, n \text{ fixed}\\ x<-1\leq x_0<...<x_{g-1}\leq 1, \underline{\mathbf{x}}\in[-1,1]^g.\end{array}\right. \quad [4.3]$$

Since the mapping

$$((w_0, ..., w_{g-1}), \underline{\mathbf{x}}) \to \sum_{i=0}^{g-1} \frac{(l_i(x))^2}{w_i}$$

is continuous, the optimization problem [4.3] can be solved in a two steps procedure.

Note for further use that

$$var\left(\mathcal{L}_n(\widehat{P})(x)\right) = \sigma^2 \sum_{j=0}^{g-1} l_j^2(x)/n_j. \qquad [4.4]$$

REMARK 4.1.– How should the minimization of $var\left(\mathcal{L}_n(\widehat{P})(x)\right)$ be performed? A simple semi-simulation procedure would help, as follows: simulate g nodes on $[-1,1]$; optimize the frequencies n_j/n given the nodes, using Karush-Kuhn-Tucker theorem; then evaluate the variance of $\mathcal{L}_n(\widehat{P})(x)$ through the above formula [4.4]; iterate the simulation of the g nodes until you find a minimum value for. Notice that $x \to var\left(\mathcal{L}_n(\widehat{P})(x)\right)$ is a polynomial with respect to x at any step of this algorithm. What we want to emphasize here is that we should not confuse the analytic expression of $var\left(\mathcal{L}_n(\widehat{P})(x)\right)$ as given in [4.4] with the fact that it is by nature the evaluation of a polynomial function. This remark proves to be the ingredient for the effective calculation of the design through an analytic approach.

The principle is, thus, as follows: fix the vector $\underline{\mathbf{x}} := (x_0, ..., x_{g-1})'$ and look for the minimum with respect to the vector $\underline{\mathbf{w}} := (w_0, ..., w_{g-1})'$. Once the optimizer $\underline{\mathbf{w}}^*$ is obtained vary $\underline{\mathbf{x}}$ and determine the resulting minimum value of the variance for fixed $\underline{\mathbf{w}}=\underline{\mathbf{w}}^*$.

Define with fixed $\underline{\mathbf{x}}$, that which determines the elementary Lagrange polynomials l_i,

$$\begin{cases} \min_{(w_0,...,w_{g-1})\in\mathbb{R}^g} \sum_{i=0}^{g-1} \frac{(l_j(x))^2}{w_j} \\ \sum_{i=0}^{g-1} w_i = n, \, n \text{ fixed} \end{cases}. \qquad [4.5]$$

Denote $\left(w_0^*, ..., w_{g-1}^*\right)$ the solution of this problem.

The optimal design ξ^* results as the solution of the following problem, assuming in the present case that $x < -1$

$$\begin{cases} \min_{\underline{\mathbf{x}}\in\mathbb{R}^g} \sum_{i=0}^{h} \frac{(l_j(x))^2}{w_j^*} \\ x < -1 \leq x_0 < .. < x_{g-1} \leq 1 \end{cases}. \qquad [4.6]$$

STEP 1.– We handle problem [4.5].

PROPOSITION 4.1.– The solution of problem [4.5] exists and is unique. It is given by

$$w_j^* := n\frac{|l_j(x)|}{\sum_{j=0}^{g-1}|l_j(x)|}, j = 0, ..., g-1.$$

PROOF.– Applying the Karush-Kuhn-Tucker theorem (see e.g. [BAZ 06]), we obtain

$$\begin{cases} \frac{\partial}{\partial w_j}\left(\sum_{i=0}^{g-1}\frac{(l_i(x))^2}{w_i} + \lambda\left(\sum_{i=0}^{g-1} w_i - n\right)\right) = 0 \\ \lambda\left(\sum_{i=0}^{g-1} w_i - n\right) = 0 \\ \lambda \geq 0 \end{cases},$$

$$\frac{\partial}{\partial w_j}\left(\sum_{i=0,i\neq j}^{g-1}\frac{(l_i(x))^2}{w_j} + \frac{(l_j(x))^2}{w_j} + \lambda\left(\sum_{i=0.i\neq j}^{g-1} w_i + w_j - n\right)\right) = 0,$$

$$\lambda = \left(\frac{l_j(x)}{w_j}\right)^2,$$

$$w_j = \frac{|l_j(x)|}{\sqrt{\lambda}}.$$

Since $\sum_{i=0}^{g-1} w_i = n$, we obtain

$$\left(\frac{l_j(x)}{w_j}\right)^2 = \lambda = \left(\frac{\sum_{j=0}^{g-1}|l_j(x)|}{n}\right)^2.$$

Finally, we find the solution of problem [4.5], namely

$$w_j^* := n\frac{|l_j(x)|}{\sum_{j=0}^{g-1}|l_j(x)|}, j = 0, ..., g-1. \qquad [4.7]$$

□

STEP 2.– Solving problem [4.6] is more tedious and requires some technical arguments, which we will develop now.

Substituting the solution [4.7] in problem [4.6] yields

$$\begin{cases} \min_{\underline{\mathbf{x}}\in\mathbb{R}^g}\sum_{i=0}^{g-1}|l_i(x)| \\ x < -1 \leq x_0 < .. < x_{g-1} \leq 1 \end{cases}. \qquad [4.8]$$

Since $x < x_0 < ... < x_{g-1}$, we have

$$|l_i(x)| = (-1)^i l_i(x). \tag{4.9}$$

Let

$$Q(x) := \sum_{i=0}^{g-1} |l_i(x)|.$$

Note that the function

$$v \to Q(v)$$

is not a polynomial on $[-1, 1]$ since only for $|v| > 1$ it holds $|l_i(v)| = (-1)^i l_i(v)$.

Outside $[-1, 1]$ the function

$$v \to Q(v) = \sum_{j=0}^{g-1} (-1)^j l_j(v) \tag{4.10}$$

is a polynomial.

The optimal design ξ should minimize [I.1] with respect to all nodes in $[-1, 1]$ and all choices of the n_j's under the constraint $n_0 + .. + n_{g-1} = n$. For a given set of nodes, the solution of the optimization step on the frequencies changes $Var\widehat{f(x)}$ into $Q(x)$ given by

$$Q(x) = \sum_{j=0}^{g-1} |l_j(x)|$$

when $-1 \leq x \leq 1$, and which, when $x < -1$ or $x > 1$, takes the form

$$Q(x) = \sum_{j=0}^{g-1} (-1)^j l_j(x). \tag{4.11}$$

This quantity is, thus, the variance of $\widehat{f(x)}$ for fixed nodes $x_0, .., x_{g-1}$, optimized upon the number of replications.

Consider the mapping Q defined outside $[-1,1]$ by

$$v \to Q(v) = \sum_{j=0}^{g-1} (-1)^j \, l_j(v),$$

hence, a polynomial with degree at most $g-1$, and consider $Q(x)$ as the evaluation at point x of the variance of $\widehat{f(x)}$ obtained as the value of this polynomial function at point x.

We need to minimize [4.11] on the nodes, hence to identify the polynomial $v \to Q(v)$ which minimizes $Q(x)$. Continuing Q on $[-1,1]$ it is readily seen that this polynomial should have value $(-1)^j$ when evaluated on x_j, and should have derivative 0 on each of the x_j'; note further that on $[-1,1]$, Q_{g-1} may assume negative values, and that $Q(x)$ is the variance of $\widehat{f(x)}$ when x lays outside $[-1,1]$.

The only polynomial which satisfies these requirements is the best uniform polynomial approximation with degree $g-1$ of the null function on the interval $[-1,1]$, as follows from the Borel-Chebyshev theorem (theorem 1.2), hence independent on x. This proves that this optimal design is obtained in a unique way independently on the location of x, if we decide to restrict its definition through the minimization of the sup norm of a polynomial when restricted to $[-1,1]$. Note further that if we look for nodes through any other class of functions, we may loose uniformity of the extrapolation design with respect to x, and there is no alternative general theory to define those designs.

A remarkable conclusion of this choice is that the design depends on x only through its frequencies (see proposition 4.1). The support of the extrapolating design holds uniformly upon the extrapolated environmental condition.

More precisely, substituting the current variable v by any of the x_i s, we obtain

$$Q(x_i) - \mathbf{0}(x_i) = (-1)^i \; 1, i = 0, ..., g-1 \qquad [4.12]$$

where $\mathbf{0}$ denotes the null function on $[-1,1]$ and Q_{g-1} takes its maximal absolute values on the x_is.

Using the Borel–Chebyshev theorem, there exists a unique polynomial T_{g-1} with degree $g-1$ defined on $[-1,1]$ which is a solution of [4.12]. It is the best uniform approximating polynomial of the null function $\mathbf{0}$ on $[-1,1]$.

We will prove below (lemma 4.1) that this polynomial is the Chebyshev polynomial of first kind on $[-1,1]$. We now turn to its identification.

Thus, we are led to the following problem: find a polynomial $v \to T_{g-1}(v)$ with degree $g-1$ which goes through the points $\left(x_j, (-1)^j\right), 0 \leq j \leq g-1$, such that

$$\max_{v \in [-1,1]} |T_{g-1}(v) - \mathbf{0}(v)| = 1$$

and $T_{g-1}(x_j) = (-1)^j$.

Turning to the derivative of T_{g-1} we see that the x_i are defined in a unique way as the roots of T'_{g-1}, when x_i is in $[-1,1]$ for all $i = 1,..,g-1$. In a first lemma (lemma 4.1), we will show that T_{g-1} satisfies a differential equation. The second lemma (lemma 4.2) identifies the optimal nodes.

LEMMA 4.1.– The polynomial T_{g-1} is the solution of the differential equation

$$1 - T_{g-1}^2(x) = \frac{1}{(g-1)^2}(1-x^2)\left(\frac{dT_{g-1}(x)}{dx}\right)^2. \qquad [4.13]$$

PROOF.– It should hold

$$\begin{cases} T_{g-1}(x_j) = (-1)^j \\ \left(\frac{dT_{g-1}(x)}{dt}\right)_{x=x_j} = 0, \; j = 0, ..., g-1. \end{cases}$$

Note that $x_0 = -1$ and $x_{g-1} = 1$. Indeed, we know that

$$|T_{g-1}(x_0)| = |T_{g-1}(x_{g-1})| = 1.$$

Now, if x_0, x_{g-1} are maximum and minimum points of T_{g-1} we have two possibilities. Either

1) both are interior points and $\left(\frac{dT_{g-1}(x)}{dx}\right) = 0$ for $x \in \{x_0, x_{g-1}\}$;

or

2) both are frontier points and $\left(\frac{dT_{g-1}(x)}{dx}\right)$ does not exist.

Assume that x_0 and x_{g-1} are interior points in $[-1,1]$; then the polynomial $\frac{dT_{g-1}(x)}{dx}$ (which is of degree $g-2$) is the null polynomial, since it has $g+1$ roots. This implies that $T_{g-1}(x)$ is the null polynomial, a contradiction. Hence,

$$x_0 = -1, \text{ and } x_{g-1} = 1.$$

Observe that:

a) the degree of $(1 - T_{g-1}^2(x))$ equals $2\,(g-1)$ and $1 - T_{g-1}^2(x_j) = 0$, which implies that the roots of $1 - T_{g-1}^2(x)$ are $\{x_0, ..., x_{g-1}\}$;

b) since $T_{g-1}(x) = a_0 + a_1 x + ... + a_{g-1}x^{g-1}$, it follows that $\frac{dT_{g-1}(x)}{dx} = a_1 + a_2 x + ... + (g-1)\,a_{g-1}x^{g-2}$, and

$$\left(\frac{dT_{g-1}(x)}{dx}\right)^2 = \beta_1 + ... + \beta_{g-2}x^{2(g-2)-2}.$$

This implies that

$$deg\left((1-x^2)\left(\frac{dT_{g-1}(x)}{dx}\right)^2\right) \leq 2\,(g-1)\,.$$

Furthermore

$$(1-x^2)\left(\frac{dT_{g-1}(x)}{dx}\right)^2 = 0$$

when $x \in \{-1, 1\} = \{x_0, x_{g-1}\}$, and

$$T'(x) = 0 \text{ when } x \in \{x_1, ..., x_{g-2}\}.$$

From a) and b) we deduce that the polynomials $1 - T_{g-1}^2(x)$ and $(1-x^2)\left(\frac{dT_{g-1}(x)}{dx}\right)^2$ share the same roots and the same degree, hence are equal up to a constant K, i.e.

$$1 - T_{g-1}^2(x) = K(1-x^2)\left(\frac{dT_{g-1}(x)}{dx}\right)^2. \qquad [4.14]$$

We determine K. Let

$$T_{g-1}(x) = \sum_{i=0}^{g-1} a_i x^i.$$

Then the coefficient of $x^{2(g-1)}$ equals a_{g-1}^2. Also when

$$\frac{dT_{g-1}(x)}{dx} = \sum_{i=1}^{g-2} i a_i x^{i-1}$$

then

$$\left(\frac{dT_{g-1}(x)}{dx}\right)^2 = (g-1)^2 a_{g-1}^2 x^{2(g-1)-2}.$$

By [4.14] it should hold that the coefficient of greatest degree of $1 - T_{g-1}^2(x)$ equals the corresponding one of $K(1-x^2)\left(\frac{dT_{g-1}(x)}{dx}\right)^2$. This means

$$-a_{g-1}^2 x^{2(g-1)} = -Kx^2 (g-1)^2 a_{g-1}^2 x^{2(g-1)-2}$$

which yields

$$1 - T_{g-1}^2(x) = \frac{1}{(g-1)^2}(1-x^2)\left(\frac{dT_{g-1}(x)}{dx}\right)^2.$$

This is a differential equation with separable variables with solution the $T_{g-1}(x)$ which we look for. □

We now obtain the optimal nodes, solving the problem [4.6].

LEMMA 4.2.– The solution of [4.13] under the boundary conditions $T_{g-1}(-1) = T_{g-1}(+1) = 1$ is $T_{g-1}(x) = \cos((g-1)\arccos x)$.

PROOF.– Denote $y = T_{g-1}(\mathsf{v})$. For $\mathsf{v} = \widetilde{x}_{g-1} = 1$, it should hold $y = T_{g-1}(1) = 1$. Therefore at $\mathsf{v} = 1$, T_{g-1} has a maximum. This proves that for $\mathsf{v} < 1$, with v close to 1, $y = T_{g-1}(\mathsf{v})$ is increasing. Let v^* the point to the left of $\widetilde{x}_{g-1}$ such that from v^* to $\widetilde{x}_{g-1}$ the function y is always increasing, and is not increasing before v^*. Clearly, v^* is a minimizer of T_{g-1} and $y(\mathsf{v}^*) = -1$. Therefore, at $[\mathsf{v}^*, 1]$ the first derivative y' is positive. We can therefore write:

$$\sqrt{(y')^2} = |y'| = y'.$$

It follows that the equation $1 - y^2 = \frac{1}{(g-1)^2}(1-\mathsf{v}^2)(y')^2$ may be written as

$$\frac{g-1}{\sqrt{1-\mathsf{v}^2}} = \frac{y'}{\sqrt{1-y^2}}.$$

Take the primitive on all terms, namely

$$(g-1)\int \frac{d\mathsf{v}}{\sqrt{1-\mathsf{v}^2}} = \int \frac{y'}{\sqrt{1-y^2}} d\mathsf{v} + c.$$

Apply the following change of variable $\mathsf{v} = \cos\theta$ on the left hand side. It then holds

$$(g-1)\int \frac{-\sin\theta}{\sqrt{1-\cos^2\theta}}d\theta = \int \frac{dy}{\sqrt{1-y^2}} + c.$$

On the right hand side set $y(\mathsf{v}) = \cos\theta$. This yields

$$(g-1)\int \frac{-\sin\theta}{|\sin\theta|}d\theta = \int \frac{-\sin\theta}{|\sin\theta|}d\theta + c.$$

Some attention yields

$$(g-1)\arccos\mathsf{v} = \arccos y + c'.$$

Hence,

$$\cos((g-1)\arccos\mathsf{v}) = \cos(\arccos y + c').$$

Take $\mathsf{v} = 1$; then $y = 1$ so that

$$\cos((g-1)\arccos 1) = \cos(\arccos 1 + c').$$

Now $\arccos 1 = 2r\pi$, with $r = 0, 1, \ldots$. Hence $\cos(h\arccos 1) = 1$, writing $rh = r'$, with $r' \in \mathbb{Z}$. It follows that $1 = \cos(\arccos 1 + c')$. Write $\arccos 1 + c' = \beta$; this implies $1 = \cos\beta$, i.e. $\beta = 2r''\pi$, with $r'' \in \mathbb{N}$.Note that,

$$\arccos 1 + c' = 2h'\pi$$

and therefore

$$c' = 2h'\pi - \arccos 1 = 2h'\pi - 2m\pi = 2\pi(h'-m) = 2h''\pi, \text{ with } h'' \in \mathbb{Z}.$$

For the constant c' we may, therefore, consider any multiple of π, including 0, i.e. $\cos((g-1)\arccos\mathsf{v}) = \cos(\arccos y + c')$ with $c' = 0$. A solution of the initial differential equation is, therefore, given by

$$y = \cos((g-1)\arccos\mathsf{v}), \text{ for } \mathsf{v} \in [\mathsf{v}^*, 1].$$

The polynomial T_{g-1} increases from v^* to 1, it should decrease at the left of v^*. Define v^{**} the point where it starts its decline. The point v^{**} is therefore a maximum point with $y(\mathsf{v}^{**}) = 1$ and y decreases on $[\mathsf{v}^{**}, \mathsf{v}^*]$. Therefore, $y' < 0$ in $[\mathsf{v}^{**}, \mathsf{v}^*]$ and

$$-\frac{y'}{\sqrt{1-y^2}} = \frac{g-1}{\sqrt{1-\mathsf{v}^2}},$$

since $\sqrt{(y')^2} = |y'| = -y'$. Therefore,

$$\int -\frac{y'}{\sqrt{1-y^2}}dy = \int \frac{g-1}{\sqrt{1-\mathsf{v}^2}}d\mathsf{v} + c, \ \arccos y$$
$$= \cos((g-1)\arccos \mathsf{v} + c),$$

which along with a similar argument as previous yields to adopt $c = 0$. Since v^* coincides with $\widetilde{x}_{g-2}$, and $\mathsf{v}^{**} = \widetilde{x}_{g-3}$, we iterate the above arguments for all nodes until x_0; we conclude that $y = \cos((g-1)\arccos \mathsf{v})$. □

PROPOSITION 4.2.– It holds

$$x_k^* = \cos\left(\frac{k\pi}{g-1}\right), \text{ for } k = 0, ..., g-1.$$

PROOF.– The nodes x_k are the roots of y'. We obtain them through a first order differentiation. With $\theta = (g-1)\arccos \mathsf{v}$ it holds

$$y' = -\sin\theta\frac{d\theta}{d\mathsf{v}} = (\sin(h\arccos \mathsf{v}))\frac{g-1}{\sqrt{1-\mathsf{v}^2}}.$$

Note that

$$\frac{h}{\sqrt{1-\mathsf{v}^2}}\sin(h\arccos \mathsf{v}) = 0, \sin(h\arccos \mathsf{v}) = 0;$$

hence

$$h\arccos \mathsf{v} = k\pi, \arccos \mathsf{v} = \frac{k\pi}{g-1}, x_k = \cos\left(\frac{k\pi}{g-1}\right).$$

Then we obtain

$$y(x_k) = \cos((g-1)\arccos x_k)$$
$$= \cos\left((g-1)\frac{k\pi}{g-1}\right) = (-1)^k,$$

which yields

$$x_k = \cos\left(\frac{k\pi}{g-1}\right), \text{ for } k = 0, ..., g-1.$$

□

We rename the x_k in increasing order.

It follows that the optimal design is given by

$$\xi^* := \left\{ \left(\left[n \frac{|l_k(x)|}{\sum_{j=0}^{g-1} |l_j(x)|} \right]; \cos\left(\frac{k\pi}{g-1} \right) \right), \text{ for } k = 0, ..., g-1 \right\}.$$

DEFINITION 4.1.– The nodes

$$x_k = \cos\left(\frac{k\pi}{g-1} \right), \text{ for } k = 0, ..., g-1.$$

are the Chebyshev nodes on $[-1, 1]$. They are the roots of the following polynomial

$$T_{g-1}(x) := \cos\left((g-1) \arccos x\right)$$

the Chebyshev polynomial of first kind.

For complements on the Chebyshev polynomial of first kind, see Appendix 2.

EXAMPLE 4.1 (Hoel–Levine).– Consider $\mathbb{R} \to \mathbb{R}, x \mapsto P_3(x) := \sum_{j=0}^{3} \theta_j x^j$, θ_j, $j = 0; 1; 2; 3$. We intend to estimate $P_3(2)$ using $n = 52$ observations in $[-1; 1]$. The model writes as $y_i(x_j) := P_3(x) + \varepsilon_i$ with $\varepsilon_i \sim N(0; 1)$ i.i.d, for all i and j, $i = 1, ..., n_j$, $j = 0, .., 3$, $\sum_{j=0}^{3} n_j = 52$. The optimal design $\overline{\xi}^*$ is given by

$$\left(x_j = \cos\left(\frac{j\pi}{3} \right), \left[n \frac{|l_j(x)|}{\sum_{j=0}^{3} |l_j(x)|} \right] \right).$$

Therefore, $x_0 = -1, x_1 = -\frac{1}{2}, x_2 = \frac{1}{2}, x_3 = 1$ and $\xi^*(-1) = \frac{5}{52}, \xi^*\left(-\frac{1}{2}\right) = \frac{12}{52}, \xi^*\left(\frac{1}{2}\right) = \frac{20}{52}, \xi^*(1) = \frac{15}{52}$. Instead of this optimal design, consider the design with nodes **supp**$(\xi) = \left\{-1, \frac{-1}{3}, \frac{1}{3}, 1\right\}$ and weights $\xi(-1) = \xi\left(\frac{-1}{3}\right) = \xi\left(\frac{1}{3}\right) = \xi(1) = \frac{13}{52}$, it holds $var_\xi\left(\mathcal{L}_3(\widehat{P_3})(2)\right) \sim 15 > var_{\xi^*}\left(\mathcal{L}_3(\widehat{P_3})(2)\right) \sim 13$ where var_ξ and var_{ξ^*} denote respectively the variance under the corresponding design.

4.2.1. *Uniform optimal interpolation designs (according to Guest)*

Consider the uniform variance of the estimator (with respect to x in $[-1, 1]$). A natural strong criterion for optimality is defined through

$$\min_{\left\{n_j \in \mathbb{N}^*, j=0,...,g-1: \sum_{j=0}^{g-1} n_j = n\right\}} \max_{x \in [-1;1]} var\left(\mathcal{L}_n(\widehat{P})(x)\right). \qquad [4.15]$$

In this section, two goals will be reached. First we obtain the optimal design ξ^* solving [4.15]. Then, we will show that extrapolation designs are of a different natures with respect to the interpolation ones, since, as seen below,

$$var_{\xi^*}\left(\mathcal{L}_n(\widehat{P})(x)\right) \neq \min_{\xi} var\left(\mathcal{L}_n(\widehat{P})(x)\right), \text{ for } x>1$$

the minimum upon all designs ξ depends on x. Here, we consider an extrapolation design with $x > 1$.

Define the Legendre polynomials on $[-1, 1]$.

DEFINITION 4.2.– The Legendre polynomial of order $g-1$ on $[-1,1]$ is defined by

$$P_{g-1}(x) := 2^{-(g-1)} \sum_{j=0}^{\left[\frac{g-1}{2}\right]} (-1)^{g-1} \binom{g-1}{j} \binom{2(g-1-j)}{g-1} x^{g-1-2j}$$
$$= \frac{1}{2^{g-1}(g-1)!} \frac{d^{g-1}}{dx^{g-1}} \left(x^2-1\right)^{g-1}$$

REMARK 4.2.– The relation $P_{g-1}(x) := \frac{1}{2^{g-1}(g-1)!} \frac{d^{g-1}}{dx^{g-1}} \left(\left(x^2-1\right)^{g-1}\right)$ is known as Rodriguez formula; see e.g. [KOL 81].

REMARK 4.3.– Clearly, P_{g-1} has $g-1$ roots in $(-1,1)$, as seen now. The polynomial $\left(x^2-1\right)^{g-1}$ has degree $2(g-1)$, and has multiple roots at points ± 1. By the Rolle theorem, its derivative admits a root inside $(-1, 1)$. This derivative assumes also the value 0 at ± 1, since it has at least three roots in $[-1, 1]$. Apply once more the Rolle theorem to the second derivative, which takes value 0 at ± 1, since it has at least four roots. Proceeding further, the $(g-1)$th derivative has $g-1$ roots in $(-1, 1)$. Up to a constant this derivative is the Legendre polynomial P_{g-1}.

REMARK 4.4.– The value of $P_{g-1}(x)$ at $x = \pm 1$ can be obtained. It holds

$$\left(x^2-1\right)^{g-1} = (x-1)^{g-1}(x+1)^{g-1}.$$

By Leibnitz formula

$$\frac{d^{g-1}\left((x-1)^{g-1}(x+1)^{g-1}\right)}{dx^{g-1}}$$
$$= \sum_{j=0}^{g-1} \binom{g-1}{j} \frac{d^j\left((x-1)^{g-1}\right)}{dx^j} \frac{d^{g-1-j}(x+1)^{g-1}}{dx^{g-1-j}}.$$

For $j = 0, ..., g-2$, it holds

$$\left(\frac{d^j \left((x-1)^{g-1} \right)}{dx^j} \right)_{x=1} = 0$$

and

$$\left(\frac{d^{g-1} (x-1)^{g-1}}{dx^{g-1}} \right)_{x=1} = (g-1)!.$$

Henceforth,

$$\frac{d^{g-1} \left((x-1)^{g-1} (x+1)^{g-1} \right)}{dx^{g-1}} = (g-1)! 2^{g-1}.$$

This yields

$$P_{g-1}(1) = 1 \text{ and } P_{g-1}(-1) = (-1)^{g-1}. \quad [4.16]$$

We need some facts about the Lagrange elementary polynomials; denote

$$\pi(x) := \prod_{j=0}^{g-1} (x - x_j).$$

LEMMA 4.3.– It holds that

(i) $\left(\frac{d\pi(x)}{dx} \right)_{x=x_j} = 0$ for $j = 1, ..., g-2$,

iff $\left(\frac{d^2\pi(x)}{dx^2} \right)_{x=x_j} = 0$for $j = 1, ..., g-2$.

(ii) $\pi(x) = \alpha \left(x^2 - 1 \right) \phi_{g-2}(x)$,with

$$\phi_{g-2}(x) = \frac{dP_{g-1}(x)}{dx}$$

where P_{g-1} is the Legendre polynomial of order $g-1$ on $[-1, 1]$.

(iii) $\left(\frac{d}{dx} l_j(x) \right)_{x=x_j} = 0$ if and only if $\left(\frac{dP_{g-1}(x)}{dx} \right)_{x=x_j} = 0.$

PROOF.– Write

$$l_j(x) = \frac{\pi(x)}{(x - x_j)\left(\frac{d\pi(x)}{dx}\right)_{x=x_j}}.$$

We have,

$$\pi(x) = (x - x_j)\left(\frac{d\pi(x)}{dx}\right)_{x=x_j} l_j(x),$$

$$\frac{d\pi(x)}{dx} = (x - x_j)\left(\frac{d\pi(x)}{dx}\right)_{x=x_j} \frac{dl_j(x)}{dx} + \left(\frac{d\pi(x)}{dx}\right)_{x=x_j} l_j(x)$$

and

$$\frac{d^2\pi(x)}{dx^2} = \left(\frac{d\pi(x)}{dx}\right)_{x=x_j} \left\{(x - x_j)\frac{d^2 l_j(x)}{dx^2} + 2\frac{dl_j(x)}{dx}\right\}.$$

This last display proves (i).

In

$$\pi(x) := \prod_{j=0}^{g-1} (x - x_j)$$

the x_j, $j = 0, ..., g-1$, are the abscissae where the variance function is minimax. In [4.9] we have proved that the absolute value of the elementary Lagrange polynomial takes value 1, which is its maximal value, when evaluated on the nodes. Hence, the variance

$$var\left(\mathcal{L}_n(\widehat{P})(x)\right) = \sum_{i=0}^{g-1} (l_j(x))^2 \frac{\sigma^2}{n_i}$$

takes its maximal values at points x_j.

Hence, $\{-1, 1\} \subset \{x_j, j = 0, ..., g-1\}$ and the remaining $g-2$ x_j are points of maximal value of the variance inside $(-1; 1)$. Write the polynomial $\pi(x)$ as

$$\pi(x) = \alpha\left(x^2 - 1\right)\phi_{g-2}(x),$$

where,

$$\alpha\phi_{g-2}(x) := \prod_{j=0,\, j\neq -1;1}^{g-1} (x - x_j).$$

The polynomial ϕ_{g-2}, with degree $g-2$ is determined through the conditions

$$\left(\frac{d^2\pi(x)}{dx^2}\right)_{x=x_j} = 0 \text{ for } j = 1, ..., g-2.$$

Since

$$\frac{d\pi(x)}{dx} = 2\alpha x\phi_{g-2}(x) + \alpha\left(x^2-1\right)\frac{d\left(\phi_{g-2}(x)\right)}{dx}$$

and

$$\frac{d^2\pi(x)}{dx^2} = 2\alpha\phi_{g-2}(x) + 4\alpha x\frac{d\left(\phi_{g-2}(x)\right)}{dx} + \alpha\left(x^2-1\right)\frac{d^2\left(\phi_{g-2}(x)\right)}{dx^2}$$

those conditions amount to the system

$$\begin{cases} 0 = 2\alpha\phi_{g-2}(x_1) + 4\alpha x_1\left(\frac{d(\phi_{g-2}(x))}{dx}\right)_{x=x_1} + \alpha\left(x_1^2-1\right)\left(\frac{d^2(\phi_{g-2}(x))}{dx^2}\right)_{x=x_1} \\ \cdots\cdots\cdots\cdots\cdots\cdots\cdots\cdots\cdots\cdots\cdots\cdots \\ 0 = 2\alpha\phi_{g-2}(x_j) + 4\alpha x_j\left(\frac{d(\phi_{g-2}(x))}{dx}\right)_{x=x_j} + \alpha\left(x_j^2-1\right)\left(\frac{d^2(\phi_{g-2}(x))}{dx^2}\right)_{x=x_j} \\ \cdots\cdots\cdots\cdots\cdots\cdots\cdots\cdots\cdots\cdots\cdots\cdots \\ 0 = 2\alpha\phi_{g-2}(x_{g-2}) + 4\alpha x_{g-2}\left(\frac{d(\phi_{g-2}(x))}{dx}\right)_{x=x_{g-2}} + \alpha\left(x_{g-2}^2-1\right)\left(\frac{d^2(\phi_{g-2}(x))}{dx^2}\right)_{x=x_{g-2}} \end{cases}.$$

Now, the derivative of the Legendre polynomial P_{g-1} is precisely the solution of this system (see [GUE 58]). Hence,

$$\phi_{g-2}(x) = \frac{dP_{g-1}(x)}{dx}.$$

This closes the proof of (ii).

We prove (iii). It holds

$$l_j(x) = \frac{\pi(x)}{(x-x_j)\left(\frac{d\pi(x)}{dx}\right)_{x_j}}.$$

Also

$$\frac{d}{dx}l_j(x) = \frac{\alpha(x^2-1)\frac{d^2}{dx^2}P_{g-1}(x) + 2\alpha x\frac{d}{dx}P_{g-1}(x)}{(x-x_j)K} - \frac{\alpha(x^2-1)\frac{d}{dx}P_{g-1}(x)}{(x-x_j)^2K^2}$$

for some constant K. When $x = x_j$ then (iii) follows. □

We now obtain the optimal design.

PROPOSITION 4.3.– The nodes of the optimal design ξ^* are the $g-2$ solutions of the equation

$$\frac{d}{dx}P_{g-1}\left(x\right)=0$$

and $-1, 1$. The optimal frequencies are all equal, and are defined by the relation

$$n_j=\frac{g}{n}.$$

PROOF.– Keeping the notation l_j for the elementary Lagrange polynomials, we have

$$var\left(\mathcal{L}_n(\widehat{P})(x)\right)=\sum_{j=0}^{g-1}l_j^2\left(x\right)\frac{\sigma^2}{n_j}.$$

Since $\frac{\sigma^2}{n_j}>0$, any $\frac{\sigma^2}{n_j}$ should be minimal to make the sum minimal. Hence, $\left(n_0^*,...,n_{g-1}^*\right)$ should solve

$$\begin{cases} min_{(n_0,...,n_{g-1})}\left(\frac{1}{n_0}+...+\frac{1}{n_{g-1}}\right) \\ \sum_{j=0}^{g-1}n_j=n \end{cases}. \qquad [4.17]$$

Hence,

$$n_j^*=\frac{g}{n}.$$

The polynomial $\sum_{j=0}^{g-1}l_{x_j}^2\left(x\right)\frac{\sigma^2}{n_j}$ has degree $2g-2$, and has $g-1$ roots of order 2. This function is a decreasing function of x on $(-\infty,-1)$ and an increasing function of x on $(1,+\infty)$; the points -1 and 1 are, therefore, points of local maximal value of the variance. The variance has, therefore, g local extrema in $[-1,1]$. Hence, there exist $g-2$ local extrema for the variance inside $(-1,1)$; they lie between the roots of $var\left(\mathcal{L}_n(\widehat{P})(x)\right)$. These extrema are maxima, since the variance is a sum of squares and takes value zero $g-2$ times.

We suppose that we have those points at hand; call them x_j, $j=0,...,g-1$. On those points, x_j the function $var\left(\mathcal{L}_n(\widehat{P})(x)\right)$ takes the value

$$\sum_{i=0}^{g-1}l_i^2\left(x_j\right)\frac{\sigma^2}{n_i}=\frac{\sigma^2}{n_j}$$

with

$$l_i(x) := \prod_{j=0, j \neq i}^{g-1} \frac{x - x_j}{x_i - x_j}.$$

The function $x \to l_j^2(x)$ takes its maximal value for $x = x_j$, with $l_j^2(x_j) = 1$ independently on j.

Recall that by formula [4.2]

$$\max_{x \in [-1;1]} var\left(\mathcal{L}_n(\widehat{P})(x)\right) = \max_{x \in [-1;1]} \sum_{j=0}^{g-1} l_j^2(x) \frac{\sigma^2}{n_j}.$$

The principle leading to the optimal design should now be made precise. The largest variance of $\left(\mathcal{L}_n(\widehat{P})(x)\right)$ should be attained on the points of measurements, to be able to control this variance. Consider two nodes $\widetilde{x}_i$ and $\widetilde{x}_k$. Then,

$$var\left(\mathcal{L}_n(\widehat{P})(x_i)\right) = var\left(\mathcal{L}_n(\widehat{P})(x_k)\right) = \frac{\sigma^2 g}{n}.$$

Hence,

$$\max_{x \in [-1;1]} var\left(\mathcal{L}_n(\widehat{P})(x)\right) = \frac{\sigma^2 g}{n}.$$

The nodes should hence be the points of maximal value of the variance, which equals $\frac{\sigma^2 g}{n}$.

The first derivative of $var\left(\mathcal{L}_n(\widehat{P})(x)\right)$ writes

$$\frac{d}{dx} var\left(\mathcal{L}_n(\widehat{P})(x)\right) = \frac{2g}{n} \sum_{j=0}^{g-1} l_j(x) \frac{d}{dx} l_j(x).$$

It follows that finding the $g-2$ internal nodes x_j's results in finding the solutions of the equation

$$\frac{d}{dx} var\left(\mathcal{L}_n(\widehat{P})(x)\right)_{x=x_j} = 0 \qquad [4.18]$$

which by the above argument turns out to solve

$$\left(\sum_{j=0}^{g-1} l_j\left(x\right)\frac{d}{dx}l_j\left(x\right)\right)_{x=x_j}=0$$

which yields, since

$$l_i\left(x_j\right)=\delta_{i,j}$$

$$\left(\frac{d}{dx}l_j\left(x\right)\right)_{x=x_j}=0 \text{ for all } j=1,..,g-2.$$

This is a system of $g-2$ equations in the $g-2$ variables $x_1,..,x_{g-2}$. This system has precisely $g-2$ solutions, solving [4.18].

Apply lemma 4.3 (iii) to conclude.

We now characterize the performance of the optimal design ξ^* through the evaluation of the minimax variance [4.15].

LEMMA 4.4.– The Legendre polynomial P_{g-1} is a solution of the following differential equation (so-called Legendre equation)

$$\left(1-x^2\right)\frac{d^2f\left(x\right)}{dx^2}-2x\frac{df\left(x\right)}{dx}+g\left(g-1\right)f\left(x\right)=0$$

i.e.

$$\frac{d}{dx}\left(\left(x^2-1\right)\frac{d}{dx}f(x)\right)=g(g-1)f(x). \qquad [4.19]$$

PROOF.– For an analytic function f on some open subset D in $\mathbb{C}$, by Cauchy formula, it holds

$$f^{(g-1)}\left(x\right)=\frac{(g-1)!}{2\pi i}\int_\gamma\frac{f\left(x\right)}{\left(z-x\right)^g}dz$$

where x is an interior point in D and γ is a regular circuit in D with x in its interior. The variable z runs on γ in the positive sense. Apply this formula to the analytic function

$$f\left(z\right)=\left(z^2-1\right)^{g-1}, g=0,1,2,....$$

By the Rodriguez formula we obtain the following relation, known as Schläfli formula (see e.g. [SAN 91])

$$P_{g-1}(x) = \frac{1}{2\pi i}\int_{\gamma} \frac{(z^2 - x)^{g-1}}{2^{g-1}(z-x)^g}dz.$$

Substituting now f by P_{g-1} in Legendre equation and applying the above formula, we obtain

$$\begin{aligned}&(1-x^2)\frac{d^2P_{g-1}}{dx^2} - 2x\frac{dP_{g-1}}{dx} + g(g-1)P \\ &= \frac{g}{2^{g-1}2\pi i}\int_{\gamma}\left(\frac{d}{dz}\left(\frac{(z^2-1)^g}{(z-x)^{g+1}}\right)\right)dz.\end{aligned}$$

Now

$$\frac{g}{2^{g-1}2\pi i}\int_{\gamma}\left(\frac{d}{dz}\left(\frac{(z^2-1)^g}{(z-x)^{g+1}}\right)\right)dz = 0.$$

This can be written through

$$\frac{d}{dx}\left((x^2-1)\frac{dP_{g-1}(x)}{dx}\right) = g(g-1)P_{g-1}(x).$$

Indeed

$$\frac{d}{dx}\left((x^2-1)\frac{dP_{g-1}(x)}{dx}\right) = 2x\frac{dP_{g-1}(x)}{dx} + (x^2-1)\frac{d^2P_{g-1}}{dx^2}$$

and therefore

$$\frac{d}{dx}\left((x^2-1)\frac{dP_{g-1}(x)}{dx}\right) = g(g-1)P_{g-1}(x)$$

which is

$$2x\frac{dP_{g-1}(x)}{dx} + (x^2-1)\frac{d^2P_{g-1}}{dx^2} - g(g-1)P_{g-1}(x) = 0$$

which proves the claim.

We evaluate the variance of the design of Guest.

We now turn back to the points where the variance assumes its maximal values. It holds

$$\pi(x) = \alpha\left(x^2 - 1\right)\frac{dP_{g-1}(x)}{dx}.$$

Hence

$$\begin{aligned}\frac{d\pi(x)}{dx} &= \frac{d}{dx}\left(\alpha\left(x^2-1\right)\frac{dP_{g-1}(x)}{dx}\right)\\ &= \alpha g(g-1)P_{g-1}(x)\end{aligned}$$

by lemma 4.4.

Therefore

$$\frac{d^2\pi(x)}{dx^2} = \alpha g(g-1)\frac{dP_{g-1}(x)}{dx}.$$

We evaluate the minimax variance, which we denote by var_{ξ^*}.

It holds

$$var_{\xi^*}\left(\mathcal{L}_n(\widehat{P})(x)\right) = \sum_{j=0}^{g-1} l_j^2(x)\frac{\sigma^2}{n_j}$$

$$= \sum_{j=0}^{g-1}\left(\frac{\pi(x)}{(x-x_j)\left(\frac{d\pi(x)}{dx}\right)_{x_j}}\right)^2\frac{n\sigma^2}{g}$$

$$= \left(\left(x^2-1\right)\frac{dP_{g-1}(x)}{dx}\right)^2\frac{n\sigma^2}{\alpha g^3(g-1)^2}\sum_{j=0}^{g-1}\left(\frac{1}{(x-x_j)P_{g-1}(x_j)}\right)^2.$$

Making use of Lobatto formula and after some calculus (see [HIL 56]), we obtain

$$var_{\xi^*}\left(\mathcal{L}_n(\widehat{P})(x)\right) = \left(1+\frac{x^2-1}{g(g-1)}\left(\frac{d^2P_{g-1}(x)}{dx^2}\right)^2\right)\frac{n\sigma^2}{g}.$$

In the extrapolation zone, namely for large $|x|>1$

$$\frac{dP_{g-1}(x)}{dx} \sim (g-1)\frac{(2(g-1))!}{2^{g-1}\left((g-1)!\right)^2}x^{g-2}.$$

In the extrapolation zone this yields to the approximation

$$var_{\xi^*}\left(\mathcal{L}_n(\widehat{P})(x)\right) \sim (g-1)\left(\frac{(2\,(g-1)!)}{2^{g-1}\left((g-1)!\right)^2}\right)^2 x^{2(g-1)}\frac{\sigma^2}{g-1}.$$

Considering the points $x_0 = -1, x_{g-1} = 1$ which are also points of maximum variance we see that the maximal variance cannot exceed $\frac{n}{g}\sigma^2$.

We have obtained the optimal minimax design in the interpolation range.

We now prove that this design is not suitable for extrapolation.

4.2.2. *The interplay between the Hoel–Levine and the Guest designs*

By

$$var_{\xi}\left(\mathcal{L}_n(\widehat{P})(x)\right) = \sum_{j=0}^{g-1} l_j^2(x)\frac{\sigma^2}{n_j}$$

the variance of $\mathcal{L}_n(\widehat{P})(x)$, say $var_{\xi}\left(\mathcal{L}_n(\widehat{P})(x)\right)$ is an increasing function of x for $x > 1$ for any design ξ since the mapping $x \to l_j^2(x)$ increases for $x \geq 1$. It follows that for any $c > 1$ the Hoel–Levine design ξ_c is the minimax optimal extrapolation design on $(1, c)$; namely it solves

$$\min_{\xi \in \mathcal{M}^*_{[-1,1]}} \max_{x \in (1,c]} var_{\xi}\left(\mathcal{L}_n(\widehat{P})(x)\right)$$

where $\mathcal{M}^*_{[-1,1]}$ designates the class of all discrete probability measures with finite support in $[-1, 1]$. However there is no reason that ξ_c be minimax optimal on whole $[-1, c]$ since it might not solve

$$\min_{\xi \in \mathcal{M}^*_{[-1,1]}} \max_{x \in [-1,c]} var_{\xi}\left(\mathcal{L}_n(\widehat{P})(x)\right).$$

We consider the optimal minimax design on $[-1, c]$ with $c > 1$ and discuss its existence and properties.

On $[-1, 1]$ the optimal minimax design is Guest's design. We will prove (see proposition 4.6 hereafter) that this design is not minimax optimal on $[-1, c]$ with $c > 1$ for large c.

On the contrary, we prove (proposition 4.4) hereafter that the Hoel–Levine design ξ_c is minimax optimal on $[1, c]$.

Finally we prove (proposition 4.5) that there exists a unique $c_1 >> 1$ such that ξ_{c^*} is minimax optimal on $[-1, c_1]$.

PROPOSITION 4.4.– The Hoel–Levine optimal design ξ_c is minimax optimal on $[1, c]$ for $c > 1$ as proved in section 4.2 (substitute $c < -1$ by $c > 1$).

PROOF.– This is a consequence of the fact that $x \rightarrow var_{\xi_c}\left(\mathcal{L}_n(\widehat{P})(x)\right)$ is an increasing function on $[1, c]$. □

PROPOSITION 4.5.– There exists $c_1 >> 1$ such that the Hoel–Levine design ξ_{c_1} is minimax optimal on $[-1, c_1]$, i.e. it solves

$$\min_{\xi \in \mathcal{M}^*_{[-1,1]}} \max_{x \in [-1, c_1]} var_{\xi}\left(\mathcal{L}_n(\widehat{P})(x)\right).$$

PROOF.– We have seen that for $1 < x < c$, the solution provided by Hoel and Levine is minimax optimal. We now consider the case when $[1, c]$ is substituted by $[-1; c]$ with $c > 1$.

In this case the minimax optimal solution still holds as the Hoel–Levine design if c is "large enough".

Indeed let $var_{\eta}\left(\mathcal{L}_n(\widehat{P})(x)\right)$ be the variance under a design η whose support consists in the Chebyshev nodes in $[-1; 1]$. The design η at this point is not defined in a unique way, since the values of $\eta(x_j)$ is not specified.

The function

$$x \rightarrow var_{\eta}\left(\mathcal{L}_n(\widehat{P})(x)\right)$$

is continuous on $[-1, 1]$. Denote

$$v^*_{\eta} := \max_{x \in [-1,1]} var_{\eta}\left(\mathcal{L}_n(\widehat{P})(x)\right).$$

Assume that there exists some $c > 1$ which does not depend on η such that

$$v^*_{\eta} < var_{\eta}\left(\mathcal{L}_n(\widehat{P})(c)\right). \quad [4.20]$$

In such a case, it holds

$$\min_{\eta \in \mathcal{M}^*_{[-1,1]}} v^*_\eta < \min_{\eta \in \mathcal{M}^*_{[-1,1]}} var_\eta \left(\mathcal{L}_n(\widehat{P})(c) \right).$$

The minimizing measure on the right hand side of the above display is precisely the extrapolation Hoel–Levine design at c since the function $x \to var_\eta \left(\mathcal{L}_n(\widehat{P})(x) \right)$ is increasing for $x > 1$.

It remains to prove that such c satisfying [4.20] exists.

For a given c, let

$$\mathbf{R}(c) := \frac{\max_{[-1;1]} \left(var_\eta \left(\mathcal{L}_n(\widehat{P})(x) \right) \right)}{var_\eta \left(\mathcal{L}_n(\widehat{P})(c) \right)} = \frac{\max_{[-1;1]} \left(\sum_{j=0}^{g-1} \frac{l_j^2(x)}{n_j} \right)}{\sum_{j=0}^{g-1} \frac{l_j^2(c)}{n_j(c)}},$$

with

$$n_j(c) := \frac{|l_j(c)|}{\sum_{i=0}^{g-1} |l_i(c)|}$$

where the $n_j(c), 0 \leq j \leq g-1$ are the optimal frequencies of the Hoel–Levine design evaluated in $x = c$.

We intend to prove that some $c > 1$ exists for which $\mathbf{R}(c) < 1$.

If this holds then

$$\mathbf{R}(c) = \frac{\left(\sum_{i=0}^{g-1} |l_i(c)| \right) \left(\max_{[-1;1]} \left(\sum_{j=0}^{g-1} \frac{l_j^2(x)}{|l_j(c)|} \right) \right)}{\left(\sum_{i=0}^{g-1} |l_i(c)| \right) \left(\sum_{j=0}^{g-1} |l_j^2(c)| \right)} = \frac{\left(\max_{[-1;1]} \left(\sum_{j=0}^{g-1} \frac{l_j^2(x)}{|l_j(c)|} \right) \right)}{\left(\sum_{j=0}^{g-1} |l_j(c)| \right)}$$

Any of the $|l_j(c)|$ is an increasing function of c for $c > 1$; therefore $\mathbf{R}(c)$ is a decreasing function of c for $c > 1$. Since each $l_j(c) \to \infty$ as $c \to \infty$, $\mathbf{R}(c)$ will approach 0 as $c \to \infty$.

Since $l_j(c) \to 0$ as $c \to 1$, for all j, $\mathbf{R}(c)$ will become infinite as $c \to 1$. But $\mathbf{R}(c)$ is a continuous function of c for $c > 1$; consequently there will exist a unique value of c, denote by c_1, satisfying $\mathbf{R}(c_1) = 1$. For $c > c_1$, $\mathbf{R}(c) < 1$; this entails that c exists with [4.20]. □

The proof of proposition 4.5 is now complete.

REMARK 4.5.– The analytic derivation of c_1 is presented in Levine [LEV 66].

It follows from the same type of arguments as that just used to reject the possibility of a Legendre (or Guest) design for $c > 1$ that the Hoel–Levine design cannot be optimum for $c < c_1$. From continuity considerations one would expect the optimum design to gradually change from the Guest spacing and weighting to the Hoel–Levine spacing and weighting as c increases from 1 to c_1. This is still an open question.

PROPOSITION 4.6.– The Guest design ξ^* is not minimax optimal on $[-1, c]$ for any $c > 1$, which is to say that it not an optimal extrapolating design.

PROOF.– By proposition 4.4, the Hoel–Levine design on $[1, c]$ is minimax optimal for large $c > 1$. By uniqueness of the optimal design, following from the optimization problem, we deduce that Guest design cannot coincide with this design. □

4.2.3. *Confidence bound for interpolation/extrapolation designs*

Using a minimax optimal design we may produce a confidence bound for $P(x)$ at any point x in $[-1, 1]$ or for x far away from $[-1, 1]$. We thus consider two cases for the location of x. When x belongs to $[-1, 1]$ then the optimal design is the Guest one. By proposition 4.5, the Hoel–Levine design is minimax on $[-1, c_1]$ for large c_1. The minimax variance on $[-1, c_1]$ is therefore the variance of $\widehat{P(c_1)}$ since $var\left(\widehat{P(x)}\right)$ is an increasing function of the variable x for $x > 1$.

Write

$$P(x) = \sum_{j=0}^{g-1} l_j(x) P(x_j) = \mathbf{l}(x)\, \mathbf{P}(\underline{\mathbf{x}})$$

where $\underline{\mathbf{x}} := (x_0, ..., x_{g-1})'$ are the Chebyshev nodes, $\mathbf{l}(x) := (l_0(x), ..., l_{g-1}(x))'$ and

$$\mathbf{P}(\underline{\mathbf{x}}) := \begin{pmatrix} P(x_0) \\ . \\ P(x_j) \\ . \\ P(x_{g-1}) \end{pmatrix}.$$

Assume that $y_i(x_j) := P(x_j) + \varepsilon_{i,j}$ with $\varepsilon_{i,j} \sim N(0;1)$ i.i.d, for all i and j, $i = 1, ..., n_j$, $j = 0, .., g-1$, $\sum_{j=0}^{g-1} n_j = n$, where the n observations are measured on $[-1;1]$. The pointwise unbiased estimator of $P(c)$, $c > 1$, is given by

$$\widehat{P(c)} := \mathbf{l}(x)\widehat{\mathbf{P}(\underline{\mathbf{x}})}$$

where

$$\widehat{\mathbf{P}(\underline{\mathbf{x}})} = \left(\frac{1}{n_0}\sum_{i=1}^{n_0} y_i(x_0), ..., \frac{1}{n_{g-1}}\sum_{i=1}^{n_{g-1}} y_i(x_{g-1})\right).$$

Since the distribution of the $y_i(\widetilde{x}_j)'s$ is $N(P(\widetilde{x}_j), 1)$ for all $i = 1, ..., n_j$ and every $j = 0, ..., g-1$, the variance of the estimator $\widehat{P(c_1)}$ is given by

$$\begin{aligned} var\left(\widehat{P(c_1)}\right) &= var\left(\sum_{j=0}^{g-1} l_j(c_1) \frac{\sum_{i=1}^{n_j} y_i(x_j)}{n_j}\right) \\ &= \sum_{j=0}^{g-1} \frac{(l_j(c_1))^2}{n_j}. \end{aligned}$$

where the n_j are the frequencies of the Hoel–Levine design evaluated at point c_1 (which is indeed the minimax optimal design on $[-1, c_1]$ as argued above). The confidence set for $P(c_1)$ is given by

$$C_n := (\mathbf{l}(c_1))'\widehat{\mathbf{P}(\underline{\mathbf{x}})} \pm \sqrt{p_\alpha \sum_{j=0}^{g-1} \frac{(l_j(c_1))^2}{n_j}}.$$

where

$$\Pr(N(0,1) > p_\alpha) = 1 - \alpha$$

and $N(0,1)$ is a random variable distributed with a standard normal law. It holds

$$\Pr(C_n \ni P(c_1)) \geq 1 - \alpha$$

When the variance of the ε_i are unknown then it can be approximated by

$$s^2 := \frac{\sum_{j=0}^{g-1} (n_j - 1) s_j^2}{n - g - 2}$$

where

$$s_j^2 := \frac{\sum_{i=1}^{n_j} \left(y_i(x_j) - \left(\sum_{i=1}^{n_j} y_i(x_j)\right)/n_j\right)^2}{n-g-2}.$$

The confidence area for $P(x)$ becomes

$$C_n := (\mathbf{1}(x))' \widehat{\mathbf{P}(\underline{\mathbf{x}})} \pm \sqrt{q_{\alpha/2} \sum_{j=0}^{g-1} \frac{(l_j(c_1))^2}{n_j s_j}}$$

where

$$\Pr\left(|t_{g-2}| > q_{\alpha/2}\right) = 1 - \alpha$$

where t_{g-2} is a Student r.v. with $g - 2$ degrees of freedom.

4.3. An application of the Hoel–Levine design

The above discussion may be applied in more general situations including the regression models. We refer to the location/scale models, which are of broad interest. Let

$$Z = \frac{Y(x) - \mu(x)}{\sigma},$$

$(\sigma, \mu) \in \mathbb{R}^+ \times \mathbf{F}$, with (σ, μ) unknown and $\mathbf{F}$ a known class of functions. The scale parameter σ is constant w.r.t. x and Z is a r.v. which is absolutely continuous w.r.t. the Lebesgue measure. Its distribution is assumed to be known and does not depend on x.

Write

$$P(x) := \mu(x) + \sigma E(Z),$$
$$\varepsilon := \sigma Z - \sigma E(Z),$$

and therefore write the location/scale model as

$$Y(x) = P(x) + \varepsilon.$$

We now consider some examples.

EXAMPLE 4.2.– The importance of the Weibull distribution in reliability is well known. Denote T a Weibull r.v. with distribution function

$$F(t) = 1 - \exp\left(-\left(\frac{t}{\mu(x)}\right)^{\beta}\right), t \geq 0.$$

It can be written

$$\ln T = \ln \mu(x) + \frac{1}{\beta} \ln(-\ln(1 - F(T))),$$

and therefore

$$Y(x) = \ln \mu(x) + \sigma Z,$$

where we write

$$Y(x) := \ln T, \sigma := \frac{1}{\beta}, Z := \ln(-\ln(1 - F(T))).$$

The model is therefore

$$Z = \frac{Y(x) - \ln \mu(x)}{\sigma}.$$

Observe that

$$\Pr(Z > t) = e^{-e^{t}}, t > 0.$$

Thus Z is the Gumbel standard r.v.

EXAMPLE 4.3.– Write the above model defining

$$\varepsilon := \sigma Z - \sigma E(Z),$$

so that

$$Y(x) = P(x) + \varepsilon,$$

where

$$P(x) := \ln \mu(x) + \sigma E(Z).$$

EXAMPLE 4.4.– For a Gaussian r.v.$X \sim N\left(\mu\left(x\right), \sigma^2\right)$, it holds

$$Z = \frac{X - \mu\left(x\right)}{\sigma} \sim N\left(0, 1\right).$$

A regression model is clearly of the proceeding type.

EXAMPLE 4.5.– Assume that T is logistic, i.e.

$$F\left(t\right) = 1 - \left(1 + \exp\left(\frac{t - P\left(x\right)}{\beta}\right)\right)^{-1}.$$

When $\beta = 1$, we may write

$$1 - F\left(t\right) = \frac{1}{1 + \exp\left(t - P\left(x\right)\right)},$$

$$1 + \exp\left(t - P\left(x\right)\right) = \frac{1}{1 - F\left(t\right)}, \ \exp\left(t - P\left(x\right)\right) = \frac{F\left(t\right)}{1 - F\left(t\right)}.$$

It is enough to state

$$Z := \ln\frac{F\left(t\right)}{1 - F\left(t\right)}, \ \varepsilon := Z - E\left(Z\right), Y := T$$

to obtain

$$Y = P\left(v\right) + E\left(Z\right) + \varepsilon.$$

EXAMPLE 4.6.– Let F be the distribution function of the rectangular r.v. on $\left[a\left(x\right), b\left(x\right)\right]$. Then

$$F\left(t\right) := \frac{t - a\left(x\right)}{b\left(x\right) - a\left(x\right)}.$$

Defining

$$Z := F\left(t\right), \ \sigma\left(x\right) := b\left(x\right) - a\left(x\right), \varepsilon := \sigma\left(x\right)\left(Z - E\left(Z\right)\right), Y := T,$$

we obtain

$$Y = P\left(x\right) + \sigma\left(x\right)E\left(Z\right) + \varepsilon.$$

The same can be generalized to polynomials of many variables and dependent data. This can be useful in industrial statistics. For example, suppose that we have to estimate the mean lifetime of some item, which may depend upon various variables. In order to stress the system to obtain an estimate in real time it may not be sufficient to act on a single variable. When the variables are the temperature and the pressure and the amount of some ingredient, some of the values may not be accessible, resulting in danger for example. In other cases values outside a normal range may lead to a change in the model.

In these cases it may be useful to act on various variables simultaneously.

4.4. Multivariate optimal designs: a special case

We extend the results of the above sections to a bivariate setting in a reliability context; extension to multivariate similar cases is straightforward.

We consider the extrapolation problem with two variables.

Let

$$P:\mathbb{R}^2\to\mathbb{R},\mathbf{x}:=(x,y)\mapsto P(x,y):=\sum_{i_1=0}^{g_1-1}\sum_{i_2=0}^{g_2-1}a_{i_1i_2}\,x^{i_1}y^{i_2},\; a_{i_1i_2}\in\mathbb{R},$$

be a polynomial in the two variables x, y with partial degrees $g_i - 1$, $i = 1, 2$ in the variables x, y. The polynomial P has $M_1 := g_1g_2$ unknown coefficients.

In order to determine these coefficients we observe P on a finite set $\mathcal{E}$ in $\mathbb{R}^2$. The fact that $\mathcal{E}$ consists of M_1 distinct points in $\mathbb{R}^2$ is not sufficient for the estimation of the coefficients; it is necessary that these points do not belong to an algebraic curve (or algebraic hypersurface in higher dimension). Indeed the identification of a polynomial with many variables usually does not have a unique solution. For example consider n points $\{(x_i, y_i) : i = 0, ..., n-1\} \subset \mathbb{R}^2$, together with n known values of P on those points $\{P(x_i, y_i) : i = 0, ..., n-1\}$; there may not exist a unique polynomial $P(x, y)$, such that

$$P(x_i,y_i)=P(x_i,y_i)\,,\;(x_i,y_i)\in\mathcal{E}.$$

Indeed it is enough to consider the case where n distinct points are on a line in $\mathbb{R}^3$. In this case there exists an infinite number of planes $z = ax + by + c$, which contain the n points (x_i, y_i).

We will therefore assume that the M_1 points which define $\mathcal{E} \subset \mathbb{R}^2$ do not belong to an algebraic curve. This implies the existence and uniqueness of a polynomial which coincides with P on $\mathcal{E}$, with partial degree $g_1 - 1$ with respect to x and $g_2 - 1$ w.r.t. y. Denote this polynomial as $P_{\mathcal{E}}(P)(.)$.

It can be proved that $(x, y) \to P_{\mathcal{E}}(P)(x, y)$ satisfies

$$P(x,y) = P_{\mathcal{E}}(P)(x,y) = \sum_{(x_i,y_i)\in\mathcal{E}} P(x_i,y_i)\, Q_i(\mathcal{E},(x,y))$$

where the polynomials $Q_i(\mathcal{E},(x,y))$ do not depend on P. Indeed we may make $Q_i(\mathcal{E},.)$ explicit; see e.g. [JOH 82 pp. 248–251].

Consider $\mathcal{E}$ a finite subset of the compact set $S := \times_{i=1}^{2}[a_i, b_i]$. Let the points (x_i, y_i) in $\mathcal{E}$ be

$$(x_{i_j}, y_{i_j})$$

$i_j = 0, ..., g_j - 1$ and $j = 1, 2$.

Define the elementary Lagrange polynomial in two variables by

$$l_{i_1 i_2}(x,y) := l_{i_1}(x)\, l_{i_2}(y) \tag{4.21}$$

where

$$l_{i_1}(x) := \frac{\prod_{h_1\neq i_1,\, h_1=0}^{g_1-1}(x - x_{h_1})}{\prod_{h_1\neq i_1,\, h_1=0}^{g_1-1}(x_{i_1} - x_{h_1})},\; l_{i_2}(z) := \frac{\prod_{h_2\neq i_2,\, h_2=0}^{g_4-1}(z - z_{h_2})}{\prod_{h_2\neq i_2,\, h_2=0}^{g_4-1}(z_{i_2} - z_{h_2})}$$

are the elementary Lagrange polynomials with respect to the coordinates x and y.

Clearly

$$l_{i_1 i_2}(x,y) = \begin{cases} 1 \text{ if } (x,y) = (x_{i_1}, y_{i_2}) \\ 0 \text{ otherwise} \end{cases}.$$

The set

$$\{l_{i_1 i_2}(x,y) : i_j = 0, ..., g_j - 1, j = 1, 2\}$$

is a basis for the linear space of all polynomials with partial degree with respect to the coordinate x_j less or equal $g_j - 1$, for $j = 1, 2$.

The Gram matrix associated with this basis

$$\left(l_{i_1 i_2}(x,y)\right)_{i_1=0,\ldots,g_1-1,.i_2=0,\ldots,g_2-1}$$

is therefore invertible; by uniqueness we have

$$Q_i(\mathcal{E},(x,y)) = l_{i_1 i_2}(x,y).$$

Therefore

$$P_{\mathcal{E}}(P)(x,y) := \mathcal{L}(P_{\mathcal{E}}(P))(x,y)$$

where we wrote

$$\mathcal{L}(P_{\mathcal{E}}(P))(x,y) = \sum_{i_1=0}^{g_1-1}\sum_{i_2=0}^{g_2-1} P(x_{i_1}, y_{i_2})\, l_{i_1 i_2}(x,y). \qquad [4.22]$$

The above formula [4.22] holds true since the nodes (x_{i_l}, y_{i_k}) belong to a rectangle (see [JOH 82]).

The polynomial $\mathcal{L}(P_{\mathcal{E}}(P))(x,y)$ is called the bivariate Lagrange polynomial. The points in $\mathcal{E}$ are the nodes for the interpolation of P.

When P is not a polynomial but merely a function defined on S, which can be extended by continuity on an open set $\mathcal{O}$ which contains S, then the Lagrange interpolation scheme may be used as an approximation scheme on $\mathcal{O}$; see [COA 66].

By uniqueness we adopt the notation

$$P_{\mathcal{E}}(P)(x,y) = \mathcal{L}(P_{\mathcal{E}}(P))(x,y) = P(x,y).$$

We now assume the following model

$$Z := \frac{Y(x,y) - P(x,y)}{\sigma}$$

where Z is a r.v. totally known in distribution with finite expectation $E(Z)$ and finite variance η^2; the scale parameter $\sigma > 0$ is unknown and does not depend on x, y; the coefficients of $P(x,y)$, $a_{i_1 i_2} \in \mathbb{R}$, are also unknown. We will see that the optimal design does not depend on the constants σ^2 nor η^2.

Denote

$$\varepsilon(x,y) := \sigma Z - \sigma E(Z)$$

whose variance equals $\eta^2\sigma^2$.

It holds

$$P(x,y) = P(x,y) + \sigma E(Z) + \varepsilon.$$

We assume then that $P(x,y)$ can be observed only on a subset S in $\mathbb{R}^2$. In the setting of accelerated runs, this subset S is the stressed domain; it will be assumed that it is a rectangle $[a_1,b_1]\times[a_2,b_2]$ in $\mathbb{R}^2$, a choice which is achievable by the experimenter. This shape allows for an important simplification for the definition and the calculation of the optimal design.

The lexicographic order on $\mathbb{R}^2$ is defined as follows; for (x,y) and (z,t) in $\mathbb{R}^2\times\mathbb{R}^2$, $(x,y)\lesssim(z,t)$ iff $x\le z, y\le t$.

Let $\mathbf{a}:=(a_1,a_2)$ be the point in $\mathbb{R}^2$ which describes the threshold between the standard operational values of the environment and the stressed conditions. With respect to $\mathbf{a}$ the stressed region is a rectangle $[\mathbf{a},\mathbf{b}]$ north-east, with south-west corner at $\mathbf{a}$, whereas the unstressed domain is the south west quadrant U with north east corner at $\mathbf{a}$. We denote $\mathbf{u}$ a point in

$$U := \left\{(x,y)\in\mathbb{R}^2 : (x,y)\lesssim\mathbf{a}\right\}.$$

We intend to find an optimal design in order to estimate the value of the polynomial P at point $\mathbf{u}$, hence we look for $\mathcal{E}$ and for the number of observations on any of the points in $\mathcal{E}$, in such a way to make the variance of the estimate of $P(\mathbf{u})$ minimal.

Let $(x_{i_1},y_{i_2})\in S$ be a node, i.e. a stress configuration. We consider now the set of trials under this configuration. Denoting $\underline{i}:=(i_1,i_2)\in\times_{j=1}^{2}\{0,...,g_j-1\}$, we define $n(\underline{i})$ as the total number of replications of the measurement Y at point (x_{i_1},y_{i_2}). We denote $\underline{\mathbf{Y}}(\underline{i})$ the vector of these measurements; we assume that the coordinates of $\underline{\mathbf{Y}}(\underline{i})$ are ordered; this is the common procedure when looking at lifetimes of a number $n(\underline{i})$ of identical systems operating in parallel during the trial. So $\underline{\mathbf{Y}}(\underline{i}) := \left(Y_{(1)}(\underline{i}),...,Y_{(n(\underline{i}))}(\underline{i})\right)$ is an ordered sample obtained from an i.i.d. sample with size $n(\underline{i})$.

The system of equations which represents the observations is therefore

$$\begin{cases} y_{(1)}(x_{i_1},y_{i_1}) = P(x_{i_1},y_{i_1}) + \sigma E(Z) + \varepsilon_1 \\ \dots\dots\dots\dots\dots\dots\dots\dots\dots\dots\dots\dots \\ y_{(k)}(x_{i_1},y_{i_2}) = P(x_{i_1},y_{i_1}) + \sigma E(Z) + \varepsilon_k \\ \dots\dots\dots\dots\dots\dots\dots\dots\dots\dots\dots\dots \\ y_{(n(\underline{i}))}(x_{i_1},y_{i_2}) = P(x_{i_1},y_{i_1}) + \sigma E(Z) + \varepsilon_{n(\underline{i})} \end{cases} . \qquad [4.23]$$

It holds

$$\sum_{i_1=0}^{g_1-1}\sum_{i_2=0}^{g_2-1} n\left(\underline{i}\right) = M_1$$

with $M_1 \in \mathbb{N} \setminus \{0\}$ a fixed number.

We assume that the vectors of measurements $\underline{\mathbf{Y}}\left(\underline{i}\right)$ and $\underline{\mathbf{Y}}\left(\underline{h}\right)$ are independent for $\underline{i} \neq \underline{h}$.

The system [4.23] may be written as

$$\underline{\mathbf{Y}}\left(\underline{i}\right) = X\theta + \varepsilon\left(\underline{i}\right)$$

where X is the matrix with $n\left(\underline{i}\right)$ rows and 2 columns

$$X\left(\underline{i}\right) := \begin{pmatrix} 1 & E\left(Z\right) \\ \cdot & \cdot \\ 1^{n(\underline{i})} & E\left(Z\right) \end{pmatrix}$$

and θ is a column vector with two rows

$$\theta := \begin{pmatrix} P\left(x_{i_1}, y_{i_2}\right) \\ \sigma \end{pmatrix}.$$

Finally

$$\underline{\varepsilon}\left(\underline{i}\right) := \begin{pmatrix} \varepsilon_1\left(\underline{i}\right) \\ \cdot \\ \varepsilon_{n(\underline{i})}\left(\underline{i}\right) \end{pmatrix}, \underline{\mathbf{Y}}\left(\underline{i}\right) := \begin{pmatrix} y_{(1)}\left(x_{i_1}, y_{i_2}\right) \\ \cdot \\ y_{(n(\underline{i}))}\left(x_{i_1}, y_{i_2}\right) \end{pmatrix}.$$

Denote

$$\Omega^{-1}\left(\underline{i}\right) := \left(cov\left(y_{(a)}\left(\underline{i}\right), y_{(b)}\left(\underline{i}\right)\right)\right)_{a,b}^{-1}$$

which we assume to exist for all $\underline{i}$. In the above display, $\Omega^{-1}\left(\underline{i}\right)$ is a matrix of order $n(\underline{i})$; the matrix $\Omega\left(\underline{i}\right)$ is not the identity matrix of order $n\left(\underline{i}\right)$ since the vector of observations is ordered according to its coordinates.

$$\Omega^{-1}\left(\underline{i}\right) := \begin{pmatrix} \omega_{1,1} & \cdot & \omega_{1,n(\underline{i})} \\ \cdot & \cdot & \cdot \\ \omega_{n(\underline{i}),1} & \cdot & \omega_{n(\underline{i}),n(\underline{i})} \end{pmatrix}.$$

The expected value of the measurement Y at point (x_{i_1}, y_{i_2}) equals $P(x_{i_1}, y_{i_2}) + \sigma E(Z)$. Denote m_{i_1,i_2} its Generalized Least Square estimator

$$m_{\underline{i}} := m_{i_1,i_2} = \left[\left(X'(\underline{i})\,\Omega^{-1}(\underline{i})\,X(\underline{i})\right)\right]^{-1} X'(\underline{i})\,\Omega^{-1}(\underline{i})\,\underline{\mathbf{Y}}(\underline{i}).$$

This estimator is strongly consistent.

We now define the estimator outside of the nodes.

Observe that for $\mathbf{u} := (x, y)$

$$\begin{aligned}\mathcal{L}(P_{\mathcal{E}}(P))(x,y) + \sigma E(Z) &:= P(x,y) + \sigma E(Z) \\ &= \sum_{(x_i,y_i)\in\mathcal{E}} (P(x_i,y_i) + \sigma E(Z))\, l_{i_1 i_2}(x,y).\end{aligned}$$

Denote $m(\mathbf{u})$ the resulting estimator of $P(\mathbf{u}) + \sigma E(Z)$

$$\begin{aligned}m(\mathbf{u}) &:= \sum_{(x_i,y_i)\in\mathcal{E}} m_{\underline{i}}\, l_{i_1 i_2}(\mathbf{u}) \\ &= \sum_{(x_i,y_i)\in\mathcal{E}} m_{\underline{i}}\, l_{i_1}(x)\, l_{i_2}(y).\end{aligned}$$

This factorization relies on the fact that S is a rectangle. We now evaluate the variance of the unbiased estimator $m(\mathbf{u})$; by independence of the measurements on the nodes

$$\begin{aligned}Var(m(\mathbf{u})) &= \sum_{\underline{i}=(i_1,i_2)}^{2} (l_{i_1}(x)\, l_{i_2}(y))^2\, var(m_{\underline{i}}) \\ &= \sum_{\underline{i}=(i_1,i_2)} (l_{i_1}(x)\, l_{i_2}(y))^2\, G\left(\sigma^2\eta^2, \Omega^{-1}(\underline{i}), X'(\underline{i})\right),\end{aligned}$$

where

$$G\left(\sigma^2\eta^2, \Omega^{-1}(\underline{i}), X'(\underline{i})\right) := \sigma^2\eta^2 \left(X'(\underline{i})\,\Omega^{-1}(\underline{i})\,X(\underline{i})\right)^{-1} X'(\underline{i}). \quad [4.24]$$

Note that $var(\mathcal{L}(P_{\mathcal{E}}(P))(x,y)) \to 0$, for $n(\underline{i}) \to \infty$, due to the fact that the generalized lest-squares estimator is consistent under the present conditions.

The optimal design results in the solution to the following optimization problem

$$\begin{cases} \min \sum_{\underline{i}=(i_1,i_2)} \left(l_{i_1\, i_2}\left(x,y\right)\right)^2 G\left(\sigma^2\eta^2, \Omega^{-1}\left(\underline{i}\right), X'\left(\underline{i}\right)\right) \\ \mathbf{u} \in U \\ M_1 = \sum_{\underline{i}} n\left(\underline{i}\right). \end{cases}$$

where the minimization is held on all choices of the set of measurements (nodes)$\mathcal{E}$ and all frequencies $n\left(\underline{i}\right)$.

Although the problem generally has a numerical solution, in some practical cases it is possible to obtain and analytic solution.

We explore a special case.

Define

$$\Gamma := \sum_{m,u}^{n(\underline{i})} \omega_{m,u}.$$

Let $E\left(Z\right) = 0$ and the distribution of Y be symmetric around $E\left(Y\left(u\right)\right)$.

In this case, $G\left(\sigma^2\eta^2, \Omega^{-1}\left(\underline{i}\right), X'\left(\underline{i}\right)\right)$ becomes:

$$G_1 := G\left(\sigma^2\eta^2, \Omega^{-1}\left(\underline{i}\right), X'\left(\underline{i}\right)\right) = \frac{\sigma^2\eta^2}{\Gamma}$$

which depends on $\underline{i}$ through Γ (see [LLO 52] for the proof).

In some cases G_1 may be simplified as follows

$$G_2 := g\left(\sigma^2\eta^2, \Omega^{-1}\left(\underline{i}\right), X'\left(\underline{i}\right)\right) = \frac{\sigma^2\eta^2}{n\left(\underline{i}\right)}. \qquad [4.25]$$

Indeed a necessary and sufficient condition for G_2 is

$$\begin{pmatrix} 1 \\ \cdot \\ 1 \end{pmatrix}' \Omega\left(\underline{i}\right) = \begin{pmatrix} 1 \\ \cdot \\ 1 \end{pmatrix}'.$$

(see [DOW 54]).

In many cases the function G takes on the form

$$\frac{\sigma^2\eta^2}{\alpha n\left(\underline{i}\right)+\beta}\left(1+o\left(1\right)\right)$$

where α, β are constants depending on the (known) distribution of the random variable Z, extending [4.25]; see [CEL 03].

The problem of determining the optimal design becomes

$$\begin{cases} \min \sum_{(i_1,i_2)} \frac{\left(l_{(i_1,i_2)}(\mathbf{u})\right)^2}{\alpha n(\underline{i})+\beta} \\ M_1 = \sum_{\underline{i}} n\left(\underline{i}\right), n\left(\underline{i}\right) \in \mathbb{R}^+. \end{cases}$$

where the minimum holds on the choice of the nodes $\mathcal{E}$ and on the frequencies. Fix (x_{i_1}, y_{i_2}) and apply the Karush-Kuhn-Tucker theorem to

$$\begin{cases} \min \sum_{(i_1,i_2)} \frac{\left(l_{(i_1,i_2)}(\mathbf{u})\right)^2}{\alpha n(\underline{i})+\beta} \\ M_1 = \sum_{\underline{i}} n\left(\underline{i}\right), n\left(\underline{i}\right) \in \mathbb{R}^+ \end{cases}.$$

where the minimization is held on the frequencies $n\left(\underline{i}\right)$. We obtain

$$\left[n^*\left(\underline{i}\right)\right] = \frac{\left|l_{(i_1,i_2)}\left(\mathbf{u}\right)\right| \left(\alpha M_1 + \beta \prod_{j=1}^{2} g_j\right)}{\sum_{(i_1,i_2)=\underline{0}} \left|l_{(i_1,i_2)}\left(\mathbf{u}\right)\right|} - \beta.$$

Clearly, $n^*\left(\underline{i}\right)$ depends on the (x_{i_1}, y_{i_2})'s. We substitute $n^*\left(\underline{i}\right)$ in the variance formula, i.e. in

$$\sum_{\underline{i}=\underline{0}}^{\underline{l}} \left(l_{(i_1,i_2)}\left(\mathbf{u}\right)\right)^2 \frac{\sigma^2\eta^2}{\alpha n\left(\underline{i}\right)+\beta},$$

with $l := \left(g_1 - 1\right)\left(g_2 - 1\right)$ to obtain

$$Var\left(m(\mathbf{u})\right) = \frac{\sigma^2\eta^2}{\alpha M_1 + \beta\left(\prod_{j=1}^{2} g_j\right)} \left(\sum_{\underline{i}=\underline{0}}^{\underline{l}} \left|l_{(i_1,i_2)}\left(\mathbf{u}\right)\right|\right)^2.$$

Hence, the optimal design will not depend on the value of $\sigma^2\eta^2$. Optimizing with respect to the $(x_{i_1}, y_{i_2})'$ under the constraint $\mathbf{u} \in U$, yields

$$\min_{\left\{(x_{i_1}, y_{i_2})_{i_1, i_2}\right\}} \sum_{\underline{i}=\underline{0}} \left| l_{(i_1,i_2)}(\mathbf{u}) \right| .$$

This is the same as the following two problems with one variable

$$\min \sum_{i_j=0}^{g_j-1} \left| l_{(i_1,i_2)}(\mathbf{u}) \right|, j = 1, 2.$$

With $\mathbf{a} := (a_1, a_2)$, the minimization is held when $j = 1$ on the abscissas x_j all larger than a_1 and smaller than b_1 and on the ordinates y_j, all larger than a_2 and smaller than b_2 when $j = 2$, since

$$\sum_{\underline{i}=\underline{0}} \left| l_{(i_1,i_2)}(\mathbf{u}) \right| = \sum_{i_1=0}^{g_1-1} |l_{i_1}(u_1)| \sum_{i_k=0}^{g_2-1} |l_{i_2}(u_2)| \qquad [4.26]$$

with $\mathbf{u} := (u_1, u_2)$.

Now minimizing the product in [4.26] results in two independent minimizations, one for each of the two factors, under the corresponding constraint on the respective terms u_1 and u_2. It follows that the optimal design is the combination of two Hoel–Levine marginal optimal designs.

Therefore the solution coincides with the previously obtained one, namely the Hoel–Levine design of Section 2, i.e.

$$s_j^*(i_j) = \frac{a_j + b_j}{2} + \frac{b_j - a_j}{2} \cos\left(\frac{g_j - 1 - i_j}{g_j - 1}\pi\right), i_j = 0, ..., g_j - 1, \ j = 1, 2$$

with $S := [a_1, b_1] \times [a_2, b_2]$.

5

An Introduction to Extrapolation Problems Based on Observations on a Collection of Intervals

5.1. Introduction

In the present chapter we return to the problem of the polynomial extrapolation we visited in Chapter 4; here, the set of observable points is no longer an interval $[a, b]$, but a set $X := [a, b] \cup [d, e]$, with $- < a < b < d < e < +\infty$. This introduces the extrapolation problem into the class of the missing data one, or of imputation problems.

As in Chapter 4, the model is polynomial with known degree; the errors are homoscedastic and uncorrelated, with mean 0.

These questions were considered for the first time by Hoel [HOE 65a] in dimensions 1 and 2. We will now follow Spruill's approach.

Extrapolation of polynomials was studied by Keifer and Wolfowitz [KIE 64]. They assumed that a polynomial P could be observed, with random error, at points chosen in the interval $[-1, 1]$ and sought experimental designs by minimizing the maximum variance over a subinterval $[-a, a]$. Asymptotically, optimal designs were obtained as a shrank to zero. The model chosen here is slightly different in the observations that are assumed are available only on disjoint intervals $[a, b]$, $[d, e]$ and optimal designs are sought for the minimization of the variance at a single point $u \in (b, d)$. For the problem corresponding to that studied by Kiefer and Wolfowitz [KIE 64], the symmetric case $b - a = e - d$, explicit solutions are given in terms of the trigonometric functions.

To be more concrete, the model is that of a polynomial $P : \mathbb{R} \rightarrow \mathbb{R}$, which can be observed on X only, and we intend to estimate its value at point $x = u$ for some $u \in (b, d)$. It is a natural choice to consider an unbiased linear estimator; since this estimator should be accurate, we choose it as the Gauss–Markov one, hence is evaluated on a design which makes its variance as minimal as possible.

The chain of arguments is similar in nature as the one discussed in Chapter 4; as for the notation, we define

$$\|f\|_{X,\infty} := \sup_{x \in X} |f(x)|$$

the uniform norm of a real valued function f on X.

The optimal design results from a double optimization of the variance of the estimator: one to determine the optimal frequencies of observations, and another to determine the points where these observations are to be performed.

In Chapter 4, the estimator of $P \in \mathcal{P}_{g-1}$ is obtained by choosing as a basis, $\mathcal{B}$ of $\mathcal{P}_{g-1}$, the collection of the elementary Lagrange polynomials $\mathcal{B} := \{l_0, ..., l_{g-1}\}$, where

$$l_i(x) := \prod_{j \neq i,\, j=0,\ldots,g-1} \frac{x - x_j}{x_i - x_j}.$$

In this way, the support of the optimal design appears naturally in the expression of the polynomial, since it coincides with the nodes of the interpolation polynomial.

Therefore,

$$P(x) := \sum_{j=0}^{g-1} l_j(x) P(x_j). \qquad [5.1]$$

Formula [5.1] suggests a plug-in estimate of $P(u)$ through

$$\widehat{P(u)} := \sum_{j=0}^{g-1} l_j(u) \widehat{P(x_j)} = \sum_{j=0}^{g-1} l_j(u) \overline{Y(x_i)} \qquad [5.2]$$

where

$$\overline{Y(x_i)} := \frac{1}{n_i} \sum_{h=1}^{n_i} Y_h(x_i).$$

Up to the factor n, the frequencies of the design appear in the above expression as the amount of observations which build the means $\overline{Y(x_i)}$, namely the n_j.

The variance of the estimator of $P(u)$ is

$$var\left(\widehat{P(u)}\right) = \sum_{j=0}^{g-1} (l_j(u))^2 \frac{\sigma^2}{n_j}. \qquad [5.3]$$

In Chapter 4 the support of the design is obtained as the solution of the problem

$$\left(x_0^*, ..., x_{g-1}^*\right) = \arg \min_{-1 \leq x_0 < ... < x_{g-1} \leq 1} \left(\sum_{j=0}^{g-1} \left| \prod_{j=0, j\neq i}^{g-1} \frac{u - x_j}{x_i - x_j} \right| \right). \qquad [5.4]$$

This problem turns out to minimize the supremum of the distance between the null function on $[a, b]$ ($[-1, 1]$ generically) and the class of all polynomials with exact degree $g - 1$. The resulting design has support independent of u, as a consequence to this u plays role in the definition of the frequencies only. We observe that in Chapter 4, we made the choice for a polynomial solution for this minimax problem.

Furthermore, we saw in Chapter 4 that problem [5.4] has as unique solution the $g - uple$ $\left(x_0^*, ..., x_{g-1}^*\right)$ which are the extreme points of $|T_{g-1}|$, where T_{g-1} solves the following uniform approximation problem

$$\max_{x \in [-1,1]} |T_{g-1}(x)| \leq \max_{x \in [-1,1]} |P(x)|,$$

where, P is any polynomial with exact degree. Polynomial $T_{g-1}(x)$ is the Chebyshev polynomial of the first kind with degree $g - 1$.

In the present case, we minimize the variance by choosing the nodes on $[a, b] \cup [d, e]$. As in Chapter 4, the frequencies are solved by Karush-Kuhn-Tucker optimization, solving

$$\begin{cases} \min_{(w_0,...,w_{g-1}) \in \mathbb{R}^g} \sum_{i=0}^{g-1} \frac{(l_j(u))^2}{w_j} \\ \sum_{i=0}^{g-1} w_i = n, n \text{ fixed} \end{cases}.$$

We obtain

$$w_j^* := n \frac{|l_j(u)|}{\sum_{j=0}^{g-1} |l_j(u)|}, j = 0, ..., g - 1.$$

Substituting those quantities in [5.3], with $\sigma^2 = 1$ (since the design does not depend on σ^2), the determination of the support turns to be

$$\begin{cases} \min_{(x_0,\dots,x_{g-1})\in\mathbb{R}^g} \sum_{i=0}^{g-1} |l_i(u)| \\ a \le x_0 < \dots < x_l \le b < u < d \le x_{l+1} < \dots < x_g \le e \end{cases}. \qquad [5.5]$$

A similar consideration, as seen in Chapter 4, induces identification of the support of the design, resulting from a uniform approximation of the null function, now on $[a,b] \cup [d,e]$, by a polynomial with fixed degree $g-1$, to solve

$$\|P^*\|_{X,\infty} = \min_{P \in \mathbf{P}_{g-1}} \|\mathbf{0} - P\|_{X,\infty} \qquad [5.6]$$

where $\mathbf{P}_{g-1}$ is the class of all polynomials with exact degree $g-1$. We will call problem P_{g-1} the above problem [5.6].

We now turn to the present case where $X := [a,b] \cup [d,e]$.

Achieser [ACH 92] proved that among all the polynomials with degree $g-1$ and valuation $a_{g-1} = 1$, the minimax polynomial of the null function on $X = [-1,-d] \cup [d,1]$ is a multiple of any Chebyshev polynomial.

In particular, if $g > 2$ is even, and $d \in (0,1)$, then

$$P^*_{g-1}(x) = K_g T_{(g-2)/2}\left(\frac{1+d^2-2x^2}{1-d^2}\right)$$

where, $T_k(x)$ is the $k-th$ Chebyshev polynomial, for some constant K_g defined by

$$\frac{1}{K_g} = T_{(g-2)/2}\left(\frac{1+d^2}{1-d^2}\right).$$

As a consequence of the above considerations, we consider minimax approximations of the null function the class of polynomials $\mathcal{P}_{g-1}(1)$, hence introducing the constraint $P(0) = 1$. It will be proved in the end of the Chapter that the solution of this problem determines the nodes of the optimal design, as in the classical case, when X is an interval; the main fact is that the solution to this minimax constrained problem has exact degree $g-1$, as happens in the case considered by Achieser.

We are led to introduce the following definition.

DEFINITION 5.1.– Let f belong to $\mathcal{C}^{(0)}(a,e)$, c be a point in (b,d) and α be a real number. Denote $\mathbf{P}_{g-1}(f,\alpha)$ the minimization problem

$$\min\left(\|f-P\|_{X,\infty};P\in\mathbf{P}_{g-1},P(c)=\alpha\right).$$

For a constrained minimax optimization, we refer to the section "approximation of functions by polynomials with linear constraints imposed on their coefficients" in [DZY 08b].

The support of the optimal design results from the solution of problem $\mathbf{P}_{g-1}\,(\mathbf{0},1)$; see the last section of this chapter.

Existence of the solution holds as a consequence of the following theorem pertaining to the minimization of a norm in a finite dimensional linear space.

THEOREM 5.1.– Let $(F,\|.\|_F)$ be a normed linear space on $\mathbb{R}$, and V be a subspace of F with finite dimension. Then, for any f in F, there exists at least some v^* in V such that,

$$\|f-v^*\|_F=\inf_{v\in V}\|f-v\|_F\,.$$

The fact that this result proves the existence of a solution to problem $\mathbf{P}_{g-1}\,(\mathbf{0},1)$ is handled in section 5.3.1.

The uniqueness of the solution to problem $\mathbf{P}_{g-1}\,(\mathbf{0},1)$ follows from the Borel–Chebyshev theorem adapted to the present setting. This extension pertains to the characterization of best polynomial approximation for a continuous function f, defined on X, in the class of all polynomials with a degree less than or equal to $g-1$ which satisfy a constraint $P(c)=0$ for some c in (a,e). We will see that the aforementioned Borel–Chebyshev theorem solves this problem. Since this chain of arguments is interesting and relevant, we will now take the occasion to present it in a more comprehensive way, leaving to the core of the chapter the proofs of the main ingredients.

We first note that problem of approximating f in $\mathcal{P}_{g-1}$ under the constraint $P(c)=0$ has solutions, since this class of polynomials is a finite dimensional linear subspace in $\mathcal{P}_{g-1}$ restricted to (a,e), a subspace which we denote $\mathcal{P}_{g-1}(0)$. Denoting $\mathcal{P}_{g-1}(f,0)$ the minimax problem just described, it holds

THEOREM 5.2 (Spruill's extended Borel–Chebyshev theorem).– Let f belong to $\mathcal{C}^{(0)}(X)$ and let P^* be any solution to problem $\mathcal{P}_{g-1}(f,0)$

$$\min_{P\in\mathcal{P}_{g-1}(0)} \|f-P\|_{X,\infty}.$$

Then, for any of those solutions, the function $f-P^*$ has at least g equioscillation points in $X=[a,b]\cup[d,e]$.

This result is, thus, in the same vein as the classical corresponding result for the approximation of continuous functions on an interval, in terms of equioscillation points.

We first check that any solution of problem $\mathbf{P}_{g-1}(\mathbf{0},1)$ has at least g points of equioscillation. We noticed that problems $\mathbf{P}_{g-1}(\mathbf{0},1)$ and

$$\min_{P\in\mathcal{P}_{g-1}(1)} \|P\|_{X,\infty}$$

coincide. It holds.

THEOREM 5.3.– P^* is a solution of problem $\min_{P\in\mathcal{P}_{g-1}(1)} \|P\|_{X,\infty}$, if and only if, P^* has at least g equioscillation points in X.

The connection with above result is as follows:

Let P_1^* denote a solution of problem $\mathrm{P}_{g-1}(\mathbf{0},1)$, and define P_0^* as the best uniform approximation of P_1^* in $P_{g-1}(0)$. It can be proved that P_0^* solves problem $\mathrm{P}_{g-1}(P_1^*,0)$. By Spruill's Borel–Chebyshev theorem, the polynomial $P_1^*-P_0^*$ has at least g equioscillation points in X. Furthermore, problem $\mathrm{P}_{g-1}(P_1^*,0)$ has a unique solution, which is P_0^*. Now, by the above theorem, P_1^* has at least g equioscillation points.

Assume that there exists a second solution P_2^* to problem $\mathrm{P}_{g-1}(\mathbf{0},1)$. The set of all solutions to problem $\mathrm{P}_{g-1}(\mathbf{0},1)$ is convex, as can be checked easily. Let

$$Q:=\frac{P_1^*+P_2^*}{2}$$

and z denote an equioscillation point of Q; then

$$E^*:=|\mathbf{0}(z)-Q(z)|\leq\frac{1}{2}|\mathbf{0}(z)-P_1^*(z)|+\frac{1}{2}|\mathbf{0}(z)-P_2^*(z)|\leq E^*,$$

which yields $|P_1^*(z)|=|P_2^*(z)|$, as neither of those terms can exceed E^*. It follows that both of the polynomials, P_1^* and P_2^*, coincide on at least g points. Those polynomials have degree $g-1$, hence they coincide. We have proved that there exists a unique solution to problem $\mathbf{P}_{g-1}(\mathbf{0},1)$.

A proposition by de Boor and Rice (see Proposition 5.4) pertaining to the case when the solution of problem $\mathbf{P}_{g-1}$ is substituted by problem [5.7] hereunder shows that the solution of this new problem coincides with the so-called extreme point of the Dirac δ_0 function; this will be explored in this present chapter. Furthermore, and this is an essential argument, the extreme point of the δ_0 function is unique, is a polynomial, and has oscillating points which are precisely the support of the optimal design. This is the content of proposition 5.4.

This polynomial, in turn, can be evaluated numerically through a modified Remez algorithm (see section 5.7.4). This is the de Boor–Rice algorithm. We know that Spruill [SPR 86] has shown that the constraint, $P(c) = 1$, does not yield any restriction to the new problem. Spruill shows that the de Boor–Rice algorithm solves

$$\|P^*\|_{X,\infty} = \begin{cases} \min_P \|\mathbf{0}-P\|_{X,\infty} \\ P \in \mathcal{P}_{g-1}(1) := \left\{ \begin{array}{c} P \in \mathcal{P}_{g-1} : P(c) = 1, \\ \text{for some arbitrary (but fixed) } c \in (b,d) \end{array} \right\} \end{cases} \quad [5.7]$$

We call this $\mathbf{P}_{g-1}(\mathbf{0},1)$ a problem.

As shown in section 5.3.4, the existence and uniqueness to the general problem $\mathbf{P}_{g-1}(f,1)$ holds as a by-product of the study of a problem denoted by $\mathbf{P}_{g-1}(f,0)$. This problem consists, for a given function f in $\mathcal{C}^{(0)}([a,e])$, of minimizing the sup norm $\|f-P\|_{X,\infty}$, when $P \in \mathcal{P}_{g-1}(0) := \{P \in \mathcal{P}_{g-1} : P(c) = 0$ for some arbitrary (but fixed) $c \in (b,d)\}$. The function, f, is chosen as the null function defined on X. To solve problem $\mathbf{P}_{g-1}(f,0)$, Spruill generalizes the Borel–Chebyshev theorem for constrained minimax problems. He also provides analytical solutions to problem [5.7] in some specific cases. Those solutions follow some information on P^* and its derivative, following the extended Borel–Chebyshev theorem. As a justification for the study of problem $\mathbf{P}_{g-1}(f,1)$ for an arbitrary function f in $\mathcal{C}^{(0)}(X)$, we note that this question is relevant for models involved more, when P is substituted, for example, by some continuous function. This concept will be developed in the next volume.

5.2. The model, the estimator and the criterion for the choice of the design

The observations were conducted on two disjoint intervals $[a,b]$, $[d,e]$ with $[a,b] \cap [d,e] = \phi$. Denote $X := [a,b] \cup [d,e]$. The function P is defined on $\mathbb{R}$ but it can only be observed on X. We aim at the value $P(u)$ for some point $u \in (b,d)$; we have at hand a fixed number $n > 1$ of observations $\mathbf{Y} := (Y(x_1), ..., Y(x_n))'$, with $\{x_1, ..., x_n\} \subset X$, where we may get many observations at the same point x_i.

The model is as follows:

For each $i = 1, ..., n$,

$$\begin{cases} P \in \mathcal{P}_{g-1} \text{ (an unknown polynomial with known degree } g-1) \\ Y(x_i) := P(x_i) + \varepsilon_i \\ cov(Y(x_i), Y(x_j)) = \begin{cases} 0 \text{ if } i \neq j \\ \sigma^2 \text{if } i = j \end{cases} . \end{cases}$$

This homoscedastic model has uncorrelated errors.

Since a polynomial with degree $g-1$ is determined by its values at g points, $x_0, ..., x_{g-1}$ and assuming that those points are chosen in X, we may write for any $u \in \mathbb{R}$,

$$P(u) := \mathcal{L}_{g-1}(P(u)) := \sum_{i=0}^{g-1} P(x_i) l_i(u), \text{ where } l_i(u) \qquad [5.8]$$

$$:= \prod_{j=0, j\neq i}^{g-1} \frac{u - x_j}{x_i - x_j}.$$

This provides the form of the estimator of $P(u)$ for some u in (b, d). It holds, taking into account the replicates on the measurement points x_i,

$$\begin{cases} \begin{cases} Y_1(x_0) = P(x_0) + \varepsilon_1 \\ \dots\dots\dots\dots \\ Y_{n_1}(x_0) = P(x_0) + \varepsilon_{n_1} \end{cases} \\ \dots\dots\dots\dots\dots\dots \\ \begin{cases} Y_1(x_{g-1}) = P(x_{g-1}) + \varepsilon_1 \\ \dots\dots\dots\dots \\ Y_{n_{g-1}}(x_{g-1}) = P(x_{g-1}) + \varepsilon_{n_{g-1}} \end{cases} \end{cases}$$

For any $j = 1, ..., g-1$, define the least square estimator

$$\widehat{P(x_j)} := \frac{1}{n_j} \sum_{i=1}^{n_i} Y_i(x_j).$$

Clearly, for all $j = 1, ..., g-1$, $E\left(\widehat{P(x_j)}\right) = P(x_j)$.

Define the estimator of $P(u)$ now using formula [5.8]

$$\widehat{P(u)} := \mathcal{L}_{g-1}\left(\widehat{P}\right)(u) := \sum_{j=0}^{g-1} \widehat{P(x_j)} l_j(u). \qquad [5.9]$$

This estimator is unbiased for any u in $[a, e]$

$$E\left(\widehat{P(u)}\right) = \sum_{j=0}^{g-1} E\left(\widehat{P(x_j)}\right) l_j(u)$$

$$= \sum_{j=0}^{g-1} P(x_j) l_j(u) = P(u).$$

5.2.1. *Criterion for the optimal design*

For some fixed u in (b, d), we determine the optimal design, namely, the set of nodes and the relative multiplicity of trials to be performed on each of them, to minimize the variance of the estimator of the extrapolated value of $P(u)$. Henceforth, this is a local optimization problem.

The variance of the estimator $\widehat{P(u)}$ in [5.9] depends on the choice of the x_i. The variance of the estimator is the natural criterion, which henceforth, also depends on the x_i. This turns to handle the problem of finding $\left(\left(x_0^*, ..., x_{g-1}^*\right), \left(n_1^*, ..., n_{g-1}^*\right)\right) \in X^g \times \mathbb{N}^g$ such that

$$\left(\left(x_0^*, ..., x_{g-1}^*\right), \left(n_0^*, ..., n_{g-1}^*\right)\right) := \arg \min_{((x_0,...,x_{g-1}),(n_0,...,n_{g-1}))} var\left(\widehat{P(u)}\right).$$

This means that we look for g points in X each of them repeated $n_i \geq 1$ times, $i = 0, ..., g-1$, such that $\widehat{P(u)}$ has minimal variance.

The optimal discrete probability measure ξ^* is denoted by

$$\xi^*\left(x_j^*\right) := \frac{n_j^*}{n}, j = 0, ..., g-1.$$

and its support by $\left\{x_0^*, ..., x_{g-1}^*\right\}$.

The remainder of this chapter is devoted to a generalization of the Borel–Chebyshev theorem and its consequences for the determination of the optimal design. This generalization is necessary due to the fact that the optimal design is supported by the equioscillation points of the solution of the $\mathbf{P}_{g-1}(\mathbf{0}, 1)$ problem. This will also lead to the de Boor-Rice algorithm which produces a numerical solution to this problem.

5.3. A constrained Borel–Chebyshev theorem

The scope of this section is to solve the $\mathbf{P}(f,0)$ problem for some arbitrary continuous function f defined on $[a,e]$.

We first justify our interest in this problem. It is well-known (see e.g. [RIC 64], Chapter 1) that when V is a linear space with finite dimension in the normed space $(F, \|.\|)$ then, with $f \in F$ the minimization problem

$$\min_{\varphi \in V} \|f - \varphi\|$$

has at least one solution.

We also notice that this fact does not hold in the case where $F := \mathcal{C}^{(0)}([a,e])$, $\|\cdot\|_{X,\infty}$ and $V := \mathcal{P}_{g-1}(1)$, the class of all polynomials P defined on (a,e) with degrees less than or equal to $g-1$ and which satisfy $P(c) = 1$ for some fixed c in X. V is not a linear space: for (α, β) in $\mathbb{R}^2$ with $\alpha + \beta \neq 1$, with $P, Q \in \mathcal{P}_{g-1}(1)$, it holds $(\alpha P + \beta Q) \notin \mathcal{P}_{g-1}(1)$.

Some other way has to be considered to assess that the $\mathbf{P}_{g-1}(\mathbf{0},1)$ problem admits a solution. This problem will be seen to be relevant for the determination of the optimal design. We will see that this problem is intimately connected with $\mathbf{P}_{g-1}(f,0)$, for some continuous function f and apply the above general argument.

5.3.1. *Existence of solutions to the $\mathbf{P}_{g-1}(\mathbf{0},1)$ problem*

Let $\mathcal{P}_{g-2}(0)$ be the class of all polynomials P defined on (a,e) with a degree less than or equal to $g-2$ which satisfies $P(c) = 0$ for some fixed c, clearly a linear subspace $\mathcal{C}^{(0)}([a,e])$. Also, $\dim \mathcal{P}_{g-2} = g-1$. Minimizing the sup norm of $\mathbf{0}-P$ over all polynomials with an exact degree $g-1$ satisfying $P(c) = 0$ amounts to approximate the monomial x^{g-1} over $\mathcal{P}_{g-2}(0)$; Hence, the $\mathbf{P}_{g-1}(\mathbf{0},0)$ problem has at least one solution P^* in $\mathcal{P}_{g-1}(0)$; it amounts to the consideration of a solution of the aforementioned problem in $\mathcal{P}_{g-2}(0)$ and to add the monomial x^{g-1}. Now, denote Q^* in this solution. Then, the best uniform polynomial approximation of the null function with degree $g-1$ is

$$P^* := x^{g-1} - \sum_{j=0}^{g-2} a_j^* x^j.$$

Since $P^* \neq \mathbf{0}$, there exists at least one point c such that $P^*(c) \neq 0$. Hence,

$$\frac{P^*}{P^*(c)} \in \mathcal{P}_{g-1}(1).$$

Therefore, we have found a polynomial with minimal uniform norm on X, with an exact degree $g-1$, and which satisfies $P^*(c) = 1$ for some c, which can be taken in (b, d). We have proved the existence of solutions to problem $\mathbf{P}_{g-1}(\mathbf{0}, 1)$; this solution also belongs to $\mathcal{P}_{g-1}(1)$.

We will characterize these solutions via some generalization of the Borel–Chebyshev theorem (theorem 5.4).

5.3.2. *A qualitative discussion on some constrained Borel–Chebyshev theorem*

We define

$$R^*(x_i) := (f - P^*)(x_i)$$

together with

$$E_i^* = |R^*(x_i)| := \|R^*\|_{X,\infty}$$

where P^* denotes the minimax polynomial, satisfying therefore

$$P^* := \arg \min_{P \in \mathcal{P}_{g-1}(0)} \|f - P\|_{X,\infty}$$

Assume that the Borel–Chebyshev theorem holds for this case i.e. the error function R^* equioscillates on X and stands as the error function for the best approximating polynomial for any f continuous on X. This result is indeed the cornerstone of this chapter and will be proved in the next section. We will consider the values of the function R^* on its points of equioscillation on the two subintervals $[a, b]$ and $[d, e]$.

Assume that there exist, for $g \geq 1$, points in X, among which l in $[a, b]$ and $g - l$ in $[d, e]$.

The sequences of values of $R^*(x_i)$ may differ according to the fact that all points x_i may belong to one or the other of the two components $[a, b]$ or $[d, e]$, or according to the fact that the sequence in $[a, b]$ or in $[d, e]$ begins with a maximum or with a minimum, or according to the fact that l and/or g are even or odd, etc.

To describe these possible sequences, indicated by

$$a \leq x_1 < \ldots < x_l \leq b < d \leq x_{l+1} < \ldots < x_g \leq e$$

the points of equioscillation, with

$$(-1)^{j-\epsilon_1} E^*, \epsilon_1 \in \{0,1\}$$

the values assumed by the $R^*(x_j)$'s for $j = 1, ..., l$ and by

$$(-1)^{j-\epsilon_2} E^*, \epsilon_2 \in \{0,1\}$$

the values assumed by the $R^*(x_j)$'s for $j = l+1, ..., g$, show that

$$E_i^* := |R^*(x_i)| = E^*, 1 \leq i \leq g.$$

The index i denotes the node pertaining to the oscillation point.

For example, let $\epsilon_1 = 0$, $\epsilon_2 = 1$, $l = 3$ and $g = 5$. We then obtain the following sequence for $R^*(x_j)$

$$(-1)^{1-0} E_1^*, (-1)^{2-0} E_2^*, (-1)^{3-0} E_3^* \,; (-1)^{4-1} E_4^*, (-1)^{5-1} E_5^*$$

i.e.

$$R^*(x_1) = -E^*, R^*(x_2) = E^*, R^*(x_3) = -E^*;$$
$$R^*(x_4) = -E^*, R^*(x_5) = E^*.$$

The symbol ";" indicates that we skipped from $[a,b]$ to $[d,e]$. The absence of this symbol means that all the points of equioscillation belong to the closure of among the components of X. In this case, either $l = 0$ (all points belong to $[d,e]$) or $l = g$ (all points are in $[a,b]$).

We may identify various cases. Those are: four corresponding to $(\epsilon_1, l) \in \{0,1\} \times \{e,o\}$, where $l \in \{e,o\}$ means that l is even or odd. These four cases correspond to the situation when all oscillation points belong to $[a,b]$. Another four cases correspond to the reciprocal situation when all points are in $[d,e]$ (i.e. $l = 0$, $(\epsilon_2, g) \in \{0,1\} \times \{e,o\}$). Finally, 16 cases correspond to the fact that $(\epsilon_1, \epsilon_2, l, g) \in \{0,1\} \times \{0,1\} \times \{e,o\} \times \{e,o\}$. In these last cases the points are spread on X entirely. We thus have 24 cases to be considered.

5.3.3. *Borel–Chebyshev theorem on* $[a,b] \cup [d,e]$

The above discussion is based on the following result:

THEOREM 5.4 (constrained Borel–Chebyshev theorem).– Let P^* be an optimal solution to the $\mathbf{P}_{g-1}(f,0)$ problem; then we may find g points pairwise distinct, such that $a \leq x_1 < ... < x_l \leq b < d \leq x_{l+1} < ... < x_g \leq e$ and

$$(f - P^*)(x_i) = \begin{cases} (-1)^{j-\epsilon_1} E^*, j = 1, ..., l \\ (-1)^{j-\epsilon_2} E^*, j = l+1, ..., g \end{cases}$$

where ϵ_1 and ϵ_2 belong to $\{0,1\}$and where $E^* := \|f - P^*\|_{X,\infty}$.

PROOF.– The existence of P^* results from the fact that $\mathcal{P}_{g-1}(0)$ is a finite dimensional linear space. The proof will examine all possible cases and follows the same line in any of those. It adapts the proof of the Borel–Chebyshev theorem. It proceeds via contradiction, building a polynomial whose sup norm is less than the sup norm of the minimax polynomial. Assume by contradiction that R^* oscillates only $m < g$ times and $l(m-l) \neq 0$, together $R^*(x_l) = R^*(x_{l+1})$; Let $z_j, j = 1, ..., g-2$ be the roots of R^*. Then, define the polynomial

$$Q := P^* + \varepsilon\alpha (x-c)^2 \prod_{j=1,\, j\neq l} (x - z_j)$$

where $\varepsilon > 0$ is small enough, and α is chosen such that $sgn(Q(x_1)) = sgn(R^*(x_1))$. Then, Q belongs to $\mathcal{P}_{g-1}(0)$ and it satisfies

$$\|f - Q\|_{X,\infty} < \|f - P^*\|_{X,\infty},$$

a contradiction. We refer to Spruill [SPR 86, p. 47] for a complete proof of this. □

REMARK 5.1.– The equioscillation points x_i of R^* which verify

$$(f - P^*)(x_i) = \begin{cases} (-1)^{j-\epsilon_1} E^*, j = 1, ..., l \\ (-1)^{j-\epsilon_2} E^*, j = l+1, ..., g \end{cases}$$

are called proper oscillation points of P^* in Spruill [SPR 87a].

REMARK 5.2.– In the classic case, where the minimax polynomial P^* with degree $g-1$ approximates the null function on a closed and bounded interval, then the Borel–Chebyshev Theorem requires that there exist at least $g+1$ points of equioscillation for the error function. This fact yields to the solution of the g coefficients of P^* along with the value of $\|P^*\|_\infty$. In this chapter, we have a lack of unknown quantities, since one of the constraint is written as $P^*(c) = 0$. It follows that there exist at least g points of equioscillation instead of $g+1$.

5.3.4. *From the constrained Borel–Chebyshev theorem to the support of the optimal design*

The following result provides a characterization to the solution of problem $\mathbf{P}_{g-1}(\mathbf{0},1)$.

THEOREM 5.5.– a) Problem $\mathbf{P}_{g-1}(\mathbf{0},1)$ has a unique solution P^*. ; b) P^* enjoys the following properties: 1) it oscillates on at least g points in X; 2)$P^*(c) = 1$; 3) the points $x = b$ and $x = d$ are points of equioscillation and it holds $P^*(b) = P^*(d) = \|P^*\|_{X,\infty}$; 4) in at least three of the four points, a, b, d, e, the polynomial P^* assumes absolute value E^*.

PROOF.– We prove a) items 1 and 2 in b) solution to problem $\mathbf{P}_{g-1}(\mathbf{0},1)$ i.e. to the problem of finding the best uniform approximation of the null function among all polynomials with a degree less than or equal to which have value 1 at point c results from the projection of $\mathbf{0}$ on the closed convex class of polynomials satisfying the above constraints. Uniqueness results from the following arguments:

Let

$$\widetilde{P} := \arg \inf_{P \in \mathcal{P}_{g-1}(1)} \|\mathbf{0} - P\|_{X,\infty}$$

and

$$\overline{P} := \arg \inf_{P \in \mathcal{P}_{g-1}(0)} \left\|\widetilde{P} - P\right\|_{X,\infty}.$$

Both $\widetilde{P}$ and $\overline{P}$ exist. Now,

$$\widetilde{P} - \overline{P} \in \mathcal{P}_{g-1}(1)$$

and

$$\left\|\widetilde{P} - \overline{P}\right\|_{X,\infty} \leq \left\|\widetilde{P} - P\right\|_{X,\infty}$$

for all P in $\mathcal{P}_{g-1}(1)$. Hence, by taking$Q := \widetilde{P} - P$, it follows that

$$\left\|\widetilde{P} - \overline{P}\right\|_{X,\infty} = \arg \inf_{Q \in \mathcal{P}_{g-1}(1)} \|\mathbf{0} - Q\|_{X,\infty}.$$

Hence,

$$\left\|\widetilde{P} - \overline{P}\right\|_{X,\infty} = \left\|\widetilde{P}\right\|_{X,\infty}. \qquad [5.10]$$

Since, $\mathbf{0}$ solves [5.10] and $\overline{P} \in \mathcal{P}_{g-1}(0)$, $\overline{P}$ is the best uniform approximation of $\widetilde{P}$ in $\mathcal{P}_{g-1}(0)$, and, since $\widetilde{P}$ is a continuous function, it follows from theorem 5.4

that $\overline{P} = \mathbf{0}$ by uniqueness, and that $\widetilde{P} - \mathbf{0}$ oscillates properly, in the aforementioned theorem. Convexity arguments show uniqueness of $\widetilde{P}$. The set of such polynomials $\widetilde{P}$ is convex, and all have, at least, g equioscillation points. Now, a convex combination of such solutions would also have at least g such points. If all solutions do not share at least g such points, convexity of the class of solutions would fail. Now, we have proved in section 5.3.1 that any solution of minimax approximation of the null function in $\mathcal{P}_{g-1}(1)$ has exact degree $g-1$. This proves the claim.

Points 3) and 4) are discussed in point c) of section 5.4.2. □

The following result provides an explicit form of the solution P^* to problem $\mathbf{P}_{g-1}(\mathbf{0},1)$. The next section presents a swift qualitative analysis of the polynomial P^*.

5.4. Qualitative properties of the polynomial which determines the optimal nodes

5.4.1. *The linear case*

– Assume that the approximating function of the null function belongs to $\mathcal{P}_0(1)$. Since $\mathcal{P}_0(1)$ contains only constants, then the minimax polynomial is necessarily constant. Hence $P^*(x) := \mathbf{1}(x) = 1$, for all x.

– Assume that the approximating function of the null function belongs to $\mathcal{P}_1(1)$. Then, P^* is the straight line $\alpha^* x + \beta^*$ which passes through the point $(c, 1)$. Hence,

$$\beta^* = 1 - \alpha^* c$$

and therefore

$$P^*(x) = \alpha^* x + (1 - \alpha^* c).$$

From theorem 5.5, on the other hand, we know that there exist two distinct points x_1 and x_2 with

$$P^*(x_1) = (-1)^{1-\epsilon_1} E^*, \; P^*(x_2) = (-1)^{2-\epsilon_1} E^*$$

and

$$E^* := \|P^*\|_{X,\infty}.$$

Also, from the same theorem 5.5, it holds that $x_1 = b$ and $x_2 = d$, and further that

$$P^*(b) = P^*(b) = \|P^*\|_{X,\infty} := E^*.$$

Hence,

$$\begin{cases} E^* = P^*(b) = \alpha^* x_1 + (1 - \alpha^* c) \\ E^* = P^*(d) = \alpha^* x_2 + (1 - \alpha^* c) \end{cases}$$

i.e.

$$\alpha^* = 0$$

$$P^*(x) = 1, \text{ for all } x.$$

Therefore, $E^* = 1$.

5.4.2. *The general polynomial case*

Suppose now that P^* belongs to $\mathcal{P}_{g-1}(1)$ with $g \geq 3$:

a) Observe first that with $\mathcal{P}^*_{g-1}(0) := \{P^*_{g-1} : g \geq 1, P^* \text{ a solution of } \mathbf{P}_{g-1}(\mathbf{0},1)\}$, the mapping $\|.\|_{X,\infty} : \mathcal{P}^*_{g-1}(0) \to [0,\infty)$ is not increasing as g increases. Hence, $\left\|P^*_{g-1}\right\|_{X,\infty} \leq \|P^*_1\|_{X,\infty} = 1$ for all $g \geq 1$.

b) Since $P^*(b) = P^*(d) = \max_{x\in X} P^*(x) = E^* \leq 1 = P^*(c)$, we deduce that $(P^*(x))' \mid_{x=b} \neq 0$.

Clearly, we cannot have $P^*(x) = 1$ for all x in (b,d). In such a case, P^* is no more a polynomial and, further, $(P^*)'$ would have an infinite number of roots whereas an element in $\mathcal{P}_{g-1}(1)$ has at most a finite collection of roots. It follows that in a right neighborhood of $x = b$, the polynomial P^* either is increasing or is decreasing.

This, in turn, implies that the equation $(P^*)'(x) = 0$ has at least one solution in (b,d). Assume, for example, that P^* decreases on a right neighborhood of b. The graph of the polynomial P^* goes continuously through the following three points:

$$(b, E^*) \to (c, 1) \to (d, E^*)$$

with $E^* \leq 1$.

Since P^* decreases, P^* goes through a minimum at some z with $P^*(z) = m$, $|m| \geq E^*$, and then it should increase until it reaches the point $(c, 1)$. Hence, $(b, E^*) \to (z, m) \to (c, 1)$.

Therefore, from the point $(c, 1)$ it will reach the point (d, E^*). If $E^* = 1$ then the graph of P^* will increase until a maximum value $M > 1$, which will be achieved at some $w \in (c, d)$; next the graph of P^* will decrease and reach the point (d, E^*). If$E^* < 1$ then P^* may reach the point (d, E^*)directly, in decreasing way.

We may think that P^* would follow the just above paths, going through many intermediate minimum and maximum value. This does not hold, as we see now.

A first consequence of the aforementioned discussion is that $(P^*(x))' \mid_{x=b} \neq 0$ and $(P^*(x))' \mid_{x=d} \neq 0$; furthermore, the equation $(P^*(x))' = 0$ has at least one solution in (b, e).

c) Clearly, at least three points in $Fr(X)$ are absolute extreme points of P^* in X (hence relative extreme points in $\mathbb{R}$). Assume that only two of them are extreme points. Since at least g points of equioscillation belong to X, it follows then that there should be at least $g - 2$ such points in the interior of X. Therefore, these $g - 2$ points are simple roots of $(P^*)'$ (due to the Fermat condition, i.e. the root in the interior of X and the function P^* changes sign at this point). This would, in turn, imply that the equation $(P^*(x))' = 0$ has at least $(g - 2) + 1 = g - 1$ roots (one of them lies in (b, e)). This cannot hold since $g - 1 > \deg((P^*)') = g - 2$.

If, instead, three points in $Fr(X)$ are extreme points then, as before, $(P^*(x))' = 0$ has $(g - 3) + 1 = g - 2$ roots. This is compatible with the condition $g - 2 \leq \deg((P^*)') = g-2$. Clearly, also the case when all frontier points of X are oscillation points may occur, since, in this case, $(P^*)'$ has $(g - 4) + 1 = g - 3$ roots.

d) We may exclude that $(P^*(x))' = 0$ has more than one solution in (b, e). The polynomial $|P^*|$ reaches in X its relative maximal value in at least g points. Three of them are on the frontier. Hence, at least $g - 3$ of those belong to $int(X)$. Assume that in (b, e) there exist at least two roots of $(P^*)'$. Then, $(P^*)'$ has $(g - 3) + 2 = g - 1$ roots. This cannot hold since $\deg((P^*)') = g - 2$.

$(P^*(x))' \mid_{x=d} \neq 0$: on the right side of $x_{l+1} = d$, the polynomial P^* is decreasing since $P^*(x_{l+2}) = -E^*$; to the left, $[c, d)$, the polynomial decreases from $P^*(c) = 1$ to $P^*(d) = E^*$.

e) Necessarily, $g - 2 \leq \deg(P^*) \leq g - 1$, $\deg(P^*) \leq g - 1$ holds since $P^* \in \mathcal{P}_{g-1}$.

To prove that $g - 2 \leq \deg(P^*)$, assume, by contradiction, that $\deg(P^*) = g - 3$. Then, it holds $\deg((P^*)') = g - 4$. Hence, in X there exist at least $g - 3$ simple roots for $(P^*)'$. This implies that the equation $(P^*)'(x) = 0$ has $(g - 3) + 1 = g - 2$ simple roots. This implies that $\deg((P^*)')$ is less than the number of roots of $(P^*)'$, a contradiction.

We consider a simple example which illustrates the behavior of the polynomial P^*.

EXAMPLE 5.1.– Let $l = 2$ and $g = 5$ (i.e. deg $P^* = 4$). The equioscillation points are chosen as: $x_1 = a$, $x_2 = b$, $x_3 = d$, $x_4 > d$, $x_5 \in (x_4, e)$. The sequence of values for $P^*(x_i)$, $i = 1, ..., 5$ can be obtained, up to the value of E^*, starting from the fact that $P^*(b) = P^*(d) = E^*$ and using the fact that P^* alternates sign from one point to the next one. Therefore, $P^*(a) = -E^*$, $P^*(b) = E^*$; $P^*(d) = E^*$, $P^*(x_4) = -E^*$, $P^*(x_5) = E^*$. Let $P^*(x) = \alpha_0 + \alpha_1 x + \alpha_2 x^2 + \alpha_3 x^3 + \alpha_4 x^4$ with $\alpha_4 > 0$. Then, $P^*(c) = 1$ and as $x \to -\infty$ it holds for$P^*(x) \to +\infty$. For $x \to +\infty$ it holds $P^*(x) \to +\infty$. On X we have $|P^*(x)| \leq E^* \leq 1$. If $\alpha_4 < 0$ then as $x \to -\infty$ it holds that$P^*(x) \to -\infty$. For $x \to +\infty$, $P^*(x) \to -\infty$. Hence, on X the polynomial P^* has three relative maxima and two relative minima. In $x = b$ it holds $(P^*)' > 0$ and in $x = d$ and $(P^*)' < 0$. In $x = a$ it holds, for $\alpha_4 > 0$, $(P^*)' = 0$. Instead, for $\alpha_4 < 0$, it holds $(P^*)' < 0$. The other frontier point $x = e$, is not extreme for P^*. Finally, it holds $(P^*)'(x) = 0$ for $x = x_4$, x_5 and $x = a$ when $\alpha_4 > 0$. Another root of the equation $(P^*)'(x) = 0$ lays at some point $z \in (b, e)$.

5.5. Identification of the polynomial which characterizes the optimal nodes

5.5.1. *The differential equation*

The above qualitative analysis yields the relation between the polynomial P^* and its derivative. This leads to a differential equation, the solution of which provides an explicit form to P^*.

The proof is plain, and we state it partially. It lies in the analysis of the various possible sequences $P^*(x_i)$, $i = 1, ..., g$, which depend on the parity and disparity of the values l and g, together with a close look at the behavior of $(P^*)'$ and P^* on $[a, e]$ (for example, in b and in d the polynomial P^* takes value $+E^*$, in $x = c$ it holds $P^*(c) = 1$, $(P^*)'$ assumes value 0 at some point $x = z \in (b, d)$, etc.) as seen in the preceding section. Based on P^*, we consider a new polynomial whose roots are located on the equioscillation points of P^*, hence on the x_i. On those points, it holds $P^*(x_i) = \pm E^*$ and we consider the relation between $E^* - P^*$ and (P^*) since both polynomials share the same roots. We, therefore, build two polynomials, the first one built upon $E^* - P^*$ and the second one on (P^*) and both polynomials share the same degree. This provides the differential equation. The above qualitative analysis allows to classify the multiplicity of the roots of those polynomials and also the behavior of P^* in a neighborhood of any of the equioscillation points. See Spruill [SPR 86] for a complete account.

PROPOSITION 5.1.– The polynomial P^* has maximum value at some point $x = c \in (b, d)$.

PROOF.– Observe that if $g > h$

$$\left\|P_g^*\right\|_{X,\infty} \leq \|P_h^*\|_{X,\infty}.$$

According to the Weierstrass theorem, the error of the approximation decreases as the degree of the approximation increases. Furthermore,

$$\|P_2^*\|_{X,\infty} \leq 1;$$

(see section 5.4.1); therefore there exists some c in (b, d) such that

$$P_g'(c) = 0.$$

Since, for any g, $\left\|P_g^*\right\|_{X,\infty} \leq 1$, $x = c$ is necessarily a point of maximal value. Further, for $g > 2$, it holds $\left\|P_g^*\right\|_{X,\infty} < 1$. □

THEOREM 5.6.– There exist constants k, f, w, q, E^* such that P^* is solution of one of the two following differential equations:

$$1)\ \left((P^*)'\right)^2 (x-v)(x-b)(x-d)(x-f) = k\left((E^*)^2 - (P^*)^2\right)(x-c)^2 \quad [5.11]$$

where $v = a$ and $f \geq e$ or $v = e$ and $f \leq a$;

$$2)\ \left((P^*)'\right)^2 (x-a)(x-b)(x-d)(x-e)(x-f)(x-w) \quad [5.12]$$

$$= k\left((E^*)^2 - (P^*)^2\right)(x-c)^2(x-q)^2$$

where $e < q < f < w$ or $w < f < q < a$.

PROOF.– The degree of P^* is either $g-2$ or $g-1$. Assume that $\deg(P^*) = g-2$. Assume also that $g > 2$ and that g is odd, that also l is odd and therefore $g-l$ is even. Then the sequence of values taken by P^* on the equioscillation points is completely determined. For example, when $l = 3$, $g = 7$, since $P^*(b) = P^*(d) = +E^*$ we obtain the sequence E_a^*, $-E_2^*$ E_b^* ; E_d^*, $-E_5^*$, $+E_6^*$, $-E_e^*$ (where, for each i, $E_i^* = E^*$ and i refers to the point x_i with $i = a$ corresponds to $i = 1$, $i = b$ for $i = 3 = l$, $i = d$ for $i = 4 = l+1$ and $i = e$ for $i = 7 = g-l$). Since the equation $(P^*)'(x) = 0$ has at most $g - 3$ solutions in $\mathbb{R}$, and one of them lays at point $x = c$, and considering the values of the sequence $P^*(x_i)$ together with the fact that g is even, we obtain $\lim_{x\to-\infty} P^*(x) = \infty$, $\lim_{x\to+\infty} P^*(x) = -\infty$. Furthermore, at points $x = a, b, d, e$ the polynomial $(P^*)'$ has simple roots. Hence, $(E^*)^2 - (P^*)^2 = (E^* - P^*)(E^* + P^*)$ has: I) four roots on the frontier, $E^* - P^*$ assumes value 0 in $x = b, d$ and $E^* + P^*$ assumes value 0 in $x = a, e$; II) inside X, the polynomial P^*, has $g - 4$ extreme values and therefore, $(P^*)^2$ has $2(g-4)$ extreme values. Hence, in $intX$, $(E^*)^2 - (P^*)^2$ has $2(g-4)$ roots. Therefore, the polynomial $(E^*)^2 - (P^*)^2$ has, in $\mathbb{R}$, $2(g-4) + 4 = 2g - 4$ roots. It follows that the polynomial

$$Q(x) := \left((E^*)^2 - (P^*)^2\right)(x-c)^2$$

has $(2g - 4) + 2 = 2g - 2$ roots in $\mathbb{R}$. On the other hand, the derivative $(P^*)'$ has $l - 2$ roots in (a, b), $(g - l) - 2$ roots in (d, e) (excluding the frontier points where $(P^*)' \neq 0$) and a root at $x = c \in (b, d)$. So, the polynomial $(P^*)'$ has $(l - 2) + ((g - l) - 2) + 1 = g - 3$ roots. Therefore, the polynomial

$$W(x) := (P^*)^2 (x-a)(x-b)(x-d)(x-e)$$

has $2(g-3) + 1 + 1 + 1 + 1 = 2g - 2$ roots in $\mathbb{R}$. Both polynomials, Q and W, have the same degree and same roots. They are equal up to a multiplicative constant k, $kW = Q$. The same argument holds for the other cases and this achieves the proof of the theorem. □

5.5.2. *Example*

Here are some explicit solutions for the above solutions P^*.

Consider $a = -1$, $e = 1$, $b = -d$ with $d \in (0, 1)$ and $c = 0$. Hence, $P^*(0) = 1$, $(P^*)'(0) = 0$.

Clearly, when $a = -1, e = 1, b = -d$ with $d \in (0, 1)$, $c = 0$ and $g = 2h - 1$ with $h > 1$, then $P^* = P_g^*$ and the degree of P^* is $g - 1$. Write $P^*(x) := a_0^* + a_1^* x + ... + a_{g-1}^* x^{g-1}$ the solution of problem $\mathbf{P}_{g-1}(\mathbf{0},1)$. Then, $P^*(-x) := a_0^* - a_1^* x + a_2^* x^2 + ... + (-1)^{g-1} a_{g-1}^* x^{g-1}$ solves problem $\mathbf{P}_{g-1}(\mathbf{0},1)$. Since $\mathbf{P}_{g-1}(\mathbf{0},1)$ has a unique solution, it holds $P^*(x) - P^*(-x) = 0$. Therefore, $(a_1^* + a_1^*)x + (a_3^* + a_3^*)x^3 + ... + \left(a_{g-1}^* + (-1)^{g-1} a_{g-1}^*\right) x^{g-1} = 0$. The polynomial $P^*(x) - P^*(-x) = 2a_3^* x^3 + ... + 2(-1)^{g-1} a_{g-1}^* x^{g-1}$ is the null polynomial iff all its coefficients are 0. Hence, $a_3^* = a_5^* = a_7^* = ... = 0$. It follows that $P^*(x) = a_0^* + a_2^* x^2 + ... + a_{2k-2}^* x^{2k-2}$ with

$k = \frac{g+1}{2}$. Therefore $\deg(P^*) \leq 2k-2$. Also P^* has at least $g+1-2 = g-1$ roots and $P^*(0) = 1$. It follows that $\deg(P^*) \geq 2k-2$ and therefore, $\deg(P^*) = 2k-2$.

We now consider

$$\left((P^*)'\right)^2 (x-v)(x-b)(x-d)(x-f) = k\left((E^*)^2 - (P^*)^2\right)(x-c)^2$$

with $c = 0$ and $a = -1$, $e = 1$, $b = -d$ and $d \in (0,1)$, $v = a$ and $f = e$; then

$$\left((P^*)'\right)^2 (x+1)(x-d)(x+d)(x-1) = k\left((E^*)^2 - (P^*)^2\right)x^2$$

i.e.

$$\left((P^*)'\right)^2 \left(x^2-1\right)\left(x^2-d^2\right) = k\left((E^*)^2 - (P^*)^2\right)x^2.$$

Integrating, we obtain

$$P^*(x) = E^*\, S_{\frac{g}{2}}$$

where

$$S_k(x) := \cos\left((2k-2)\tan^{-1}\left(\sqrt{\frac{x^2-d^2}{1-x^2}}\right)\right) \qquad [5.13]$$

for $x \in [d, 1)$.

We determine the expression of P^* for all $x \in \mathbb{R}$ using $k = 2, 3, 4, \ldots$ using

$$\begin{cases} S_{k+1}(x) + S_{k-1}(x) = 2S_k(x)S_2(x) \text{for } k \geq 2 \\ S_1(x) = 1, \text{ for all } x \in [d,1) \\ S_2(x) = \frac{2x^2}{d^2-1} + \frac{1+d^2}{1-d^2} \end{cases},$$

which holds for $x \in [d, 1)$. Finally

$$E^* = 2\left(\left(\frac{1+d}{1-d}\right)^{\frac{g}{2}-1} + \left(\frac{1-d}{1+d}\right)^{\frac{g}{2}-1}\right)^{-1},$$

see Spruill [SPR 86].

5.6. The optimal design in favorable cases

In the preceding example some specifications, in the domain of observation, allowed for an explicit class of solutions to problem $\mathbf{P}_{g-1}(\mathbf{0},1)$. In this section, we will call the situation a favorable case when the optimal nodes are known; this is the case when the solution of the differential equation [5.11] and [5.12] is known explicitly, or when g is very small (typically $g = 3$), in which case the nodes are obtained through the constrained Borel–Chebyshev theorem directly; this case will illustrate this better. The relation between the minimax solution of problem $\mathbf{P}_{g-1}(\mathbf{0},1)$ is confronted with the optimal design, and is shown to coincide with it, in the sense that the extreme points of the minimax solution are the supporting points of the optimal design.

Spruill [SPR 86, pp. 46–47], who is referring to Achieser [ACH 92], said "even when the two intervals are not of the same length, the solutions, which are shown to be cosines of elliptic integrals, can be characterized by their oscillation properties providing a reasonable rapid method, not involving quadratures, for finding solutions numerically. Solutions to the monic minimizer of $\|Q\|_{\infty,X}$ for odd degrees are also characterized".

From now on, we choose for c a point in (b, d), where

$$(P^* (c))' = 0.$$

Such a point exists due to point b) in section 5.4.2.

Assume that $g > 2$, when $g \in \{0, 1\}$ any choice of the design is optimal.

We first prove that the solution of problem $\mathbf{P}_{g-1}(\mathbf{0},1)$ does not depend on the value of c.

PROPOSITION 5.2.– P^* does not depend on c.

PROOF.– For $g > 2$, consider $\mathbf{P}_{g-1}(\mathbf{0},1)$ as c varies. Indicate by P_j^*, $j = 1, 2, 3, ...,$ the polynomial with minimal norm on X with degree less or equal $g - 1$ which assumes value 1 in some point c_j in (b, d). Hence, P_j^* solves $\mathbf{P}_{g-1}(\mathbf{0},1)$ for $c = c_j$. Consider two such problems; the first one with solution P_1^* and the second one with solution $P_{2,*}$. The equation in the variable α, $\alpha P_{1,*}(b) = P_{2,*}(b)$ has a solution, say $\widetilde{\alpha}$. Thus, the polynomial $Q := \widetilde{\alpha} P_1^* - P_2^*$ has at least g roots in $[a, e]$ (P_1^* and P_2^* both have degree $g - 1$) and therefore, $Q \equiv \mathbf{0}$. It follows that for all x, $\alpha P_1^*(x) = P_2^*(x)$ holds. We have proved that as c varies, all solutions to problems $\mathbf{P}_{g-1}(\mathbf{0},1)$ are multiple one of the other. Therefore, the solution P^* (and the equioscilation points) do not depend upon c. □

We now prove that the support of the optimal design coincides with the equioscilation points of P^*. According to the above result, the optimal design does not depend on c.

We evaluate the variance of the estimator of $P(u)$.

$$var\left(\widehat{P_{g-1}(u)}\right) = \sum_{j=0}^{g-1} var\left(\widehat{P(x_j)}\right) l_j^2(u) \quad [5.14]$$
$$= \sigma^2 \sum_{j=0}^{g-1} \frac{l_j^2(u)}{n_j}.$$

Since σ^2 is independent of the design, we may choose $\sigma^2 = 1$.

Minimizing [5.14] with respect to $n_j, j = 0, ..., g-1$, under the constraint

$$\sum_{j=0}^{g-1} n_j = n,$$

as in Chapter 4, we obtain

$$\xi^*(x_j) = \frac{|l_j(u)|}{\sum_{j=0}^{g-1} |l_j(c)|}, j = 0, ..., g-1.$$

As usual, in the above expression the index j identifies a node $x_j \in X$, which defines the elementary Lagrange polynomial l_j. Substitute these optimal frequencies in the variance. We obtain

$$var\left(\widehat{P_{g-1}(u)}\right) = \left(\sum_{j=0}^{g-1} |l_j(u)|\right)^2.$$

We are led to the following minimization problem

$$\begin{cases} \min_{(x_0,\ldots,x_{g-1})\in\mathbb{R}^g} \sum_{j=0}^{g-1} |l_j(u)| \\ -\infty < a \le x_0 < \ldots < x_l \le b < c < d \le x_{l+1} < \ldots < x_{g-1} \le e < +\infty \\ u \in (b,e) \setminus \{c\} \end{cases} . \quad [5.15]$$

Following the version of the Borel–Chebyshev theorem (theorem 5.5) the minimization problem [5.15] is equivalent to the optimization problem $\mathbf{P}_{g-1}(\mathbf{0}, 1)$ which in turn does not depend upon the choice of c.

We assume that problem 5.14 is solved, for example, through the example of the extension of the Remez algorithm due to de Boor and Rice, which we will present in the next section. Optimal frequencies follow from the Karush-Kuhn-Tucker optimization.

5.6.1. *Some explicit optimal designs*

The case when $g = 1 = 0$ or $g = 2$ is the linear case. Then, the minimax polynomial P^* is a constant 1 and the support of the design ξ^* is, for $g = 1$, any arbitrarily chosen point in X and, for $g = 2$, any couple of distinct points in X.

When $g = 3$ the support is made of three points which, necessarily, are frontier points, since at least three of the frontier points are oscillation points of the minimax polynomial. Since both b and d are such points, it follows that $\mathbf{supp}(\xi^*) = \{a, b, d\}$ or $\mathbf{supp}(\xi^*) = \{b, d, e\}$. We have $\mathbf{supp}(\xi^*) = \{a, b, d\}$ if $b - a > e - d$ and $\mathbf{supp}(\xi^*) = \{b, d, e\}$ if $e - d > b - a$; see Spruill [SPR 87a] for details.

When $g = 4$, then the the support of ξ^* is the frontier of X.

For $g > 2$ with g even and $a = -1$, $e = 1$, $-b = d$ with $b \in (0, 1)$, it is easy to obtain the minimax polynomial using [5.14]. It holds

$$P^*(x) = E^* \, T_{\frac{g-2}{2}} \left(\frac{1 + b^2 - 2x^2}{1 - b^2} \right) \qquad [5.16]$$

where,

$$T_{\frac{g-2}{2}}(x) := \cos\left(\left(\frac{g-2}{2} \right) \arccos x \right); \qquad [5.17]$$

from [5.16] and [5.17], we have

$$E^* := \frac{1}{T_{\frac{g-2}{2}}\left(\frac{1+b^2}{1-b^2}\right)}.$$

5.7. The optimal design in the general case

It is usually impossible to obtain an explicit solution to the differential equations [5.11] and [5.12]. In this section, we will describe a numerical solution; this approach follows Spruill [SPR 87a], who considered an extension of the Remez algorithm according to de Boor and Rice [BOO 82]. This technique provides a numerical solution to extremely complex norm problems related to polynomials. The aim is, as before, to solve the variational problem stated in [5.6], which minimizes the variance of the estimate of $P(u)$ providing, in this way, the support of the optimal design. We will provide some general instructions on such problems now.

Some preliminary result is needed. It provides a equivalent statement for the best uniform approximation of a continuous function, defined on an arbitrary compact set.

We know that X is a generic compact set in $\mathbb{R}^m$ and for any f in $\mathcal{C}^{(0)}(X, \mathbb{R})$ the critical set of f

$$\mathbf{Cr}(f) := \left\{ y \in X : |f(y)| = \|f\|_{X,\infty} \right\}.$$

Then, let V be a finite dimensional subspace in $\mathcal{C}^{(0)}(X, \mathbb{R})$ generated by a basis $\{\varphi_1, ..., \varphi_k\}$.

Let

$$\mathbf{K} := \left\{ (f(y)\varphi_1(y), ..., f(y)\varphi_k(y))', y \in \mathbf{Cr}(f) \right\} \subset \mathbb{R}^k$$

and

$$\widetilde{\mathbf{K}} := conv\{\psi_1, ..., \psi_r : \psi_i \in \mathbf{K}, i = 1, ..., r, r \leq k+1\}$$

It holds

THEOREM 5.7.– With the above notation

$$\|f + \varphi\|_{X,\infty} \geq \|f\|_{X,\infty}$$

for all φ in V if and only if $\underline{\mathbf{0}} := \left(0^1, ..., 0^k\right) \in \widetilde{\mathbf{K}}$.

The rather lengthy proof of this result is mentioned to Appendix 1.

5.7.1. *The extreme points of a linear functional*

The Remez algorithm suggests an alternative approach to the optimization problem $\mathbf{P}_{g-1}(\mathbf{0}, 1)$.

We adapt the Remez algorithm to the case, where constraints on the observation range are introduced; we are led to a constrained uniform approximation problem.

de Boor and Rice [BOO 82] proposed an iterative algorithm for this class of problems; this requires a preliminary discussion on the extreme points of a linear function.

Let $V := span\{\varphi_0, ..., \varphi_{g-1}\}$ denote a finite dimensional linear space, with dimension g, included in the norm space $\left(\mathcal{C}^{(0)}(X), \|.\|_{X,\infty}\right)$ and let

$$\Phi : V \to (\mathbb{R}, |.|)$$

denote a linear mapping. By definition,

$$|||\Phi||| := \sup_{f \in F, f \neq 0} \frac{\|\Phi(f)\|_F}{\|f\|_F} = \max_{\|f\| \leq 1} |\Phi(f)|.$$

Denote

$$m := |||\Phi||| \in \mathbb{R}_+.$$

DEFINITION 5.2.– The points $f^* \in V$, such that $\|f^*\|_{X,\infty} = 1$ (points on the frontier of the unit sphere) and such that $\Phi(f^*) = |||\Phi|||$ are called the extreme points for Φ.

EXAMPLE 5.2.– Let $V := \mathcal{P}_n([a,b])$ and $\Phi : \mathcal{P}_n([a,b]) \to \mathbb{R}$,

$$f \mapsto \Phi(f) := \frac{f^{(n)}(0)}{n!}.$$

Then, the extreme points of Φ are the Chebyshev polynomials, T_n, and $m = 2^{n-1}$. See Appendix 3 (Markov inequality [A3.1]).

The next theorem provides a useful representation of the extreme points.

5.7.2. *Some results on the representation of the extreme points*

We state a general result, in the setting of a generic linear space V with dimension k; the context for the application, at hand, is $V = \mathcal{P}_{g-1}$, and hence $k = g - 1$.

THEOREM 5.8 (Rivlin).– Let $\Phi \neq \mathbf{0}$ be a real linear functional on V. Then, there exist points $y_1, ..., y_r$ of X and nonzero real numbers $\alpha_1, ..., \alpha_r$ with $r \leq k$, such that for every $f \in V$, $\Phi(f) = \sum_{i=1}^{r} \alpha_i f(y_i)$ and $|||\Phi||| = \sum_{i=1}^{r} |\alpha_i|$.

PROOF.– Let f^* be an extreme point of Φ. Assume that $\dim V = 1$, for any $f \in V$, it holds $f = cf^*$. We then have

$$\Phi(f) = \Phi(cf^*) = c\Phi(f^*) = c\,|||\Phi|||.$$

Let $y_1 \in X$ be such that $f^*(y_1) = \epsilon$, with $\epsilon \in \{-1; 1\}$. Then,

$$f(y_1) = cf^*(y_1) = \epsilon c.$$

hence $c = \epsilon f(y_1)$. Also,

$$\Phi(f) = \Phi(cf^*) = c\,|||\Phi||| = \epsilon f(y_1)\,|||\Phi|||.$$

We know $\alpha := \epsilon\,|||\Phi||| \in \mathbb{R}$, the function Φ writes

$$\Phi(f) = \alpha f.$$

The theorem is, therefore, true for $\dim V = k$ for $k = 1$. Assume now $k > 1$. Then $\dim Ker(\Phi) \leq n - 1$. Since $n > 1$, there exists at least some $f_0 \in Ker(\Phi)$, $f_0 \neq \mathbf{0}$. Hence,

$$|\Phi(f^* + f_0)| = |\Phi(f^*) + \Phi(f_0)| = |\Phi(f^*) + 0| = |||\Phi|||.$$

By the definition of the norm, $|||\Phi||| := \max_{\|h\| \leq 1} |\Phi(h)|$. With

$$h := \frac{f^* + f_0}{\|f^* + f_0\|_{X,\infty}} \in S(f; 1)$$

it holds

$$\begin{aligned} |\Phi(h)| &= \left| \Phi\left(\frac{f^* + f_0}{\|f^* + f_0\|_{X,\infty}} \right) \right| = \frac{|\Phi(f^*) + \Phi(f_0)|}{\|f^* + f_0\|_{X,\infty}} \\ &\leq \max_{\|f\| \leq 1} |\Phi(f)| = |||\Phi|||, \end{aligned}$$

i.e.

$$1 \leq \|f^* + f_0\|_{X,\infty}.$$

Furthermore, $\|f^*\|_{X,\infty} = 1$. Hence,

$$\|f^* + f_0\|_{X,\infty} \geq \| f^*\|_{X,\infty}$$

applying the preceding theorem 5.7 to the linear space $Ker\,(\Phi)$ to $g := f^*$, there exist, $r \leq k$, points $y_1, ..., y_r$ in $\mathbf{Cr}\,(f)$ and real non-negative numbers $\alpha_1, ..., \alpha_r$ such that

$$\sum_{i=1}^{r} \alpha_i f^*\,(y_i)\, f_0\,(y_i) = 0, \text{ for all } f_0 \in Ker\,(\Phi)\,.$$

If $f \in V$, then

$$f_0 := ((\Phi\,(f))\, f^* - (\Phi\,(f^*))\, f) \in Ker\,(\Phi)\,.$$

We will now prove that $\Phi\,(f_0) = 0$. It holds,

$$\begin{aligned}\Phi\,(f_0) &= \Phi\,((\Phi\,(f))\, f^* - (\Phi\,(f^*))\, f) = \Phi\,(f)\,\Phi\,(f^*) - (\Phi\,(f^*))\, f \\ &= \Phi\,(f^*)\,(\Phi\,(f) - f) = \Phi\,(f^*)\,\Phi\,(f - f) = \Phi\,(f^*)\,\Phi\,(0) \\ &= \Phi\,(f^*) \cdot 0 = 0.\end{aligned}$$

If

$$\sum_{i=1}^{r} \alpha_i f^*\,(y_i)\, f_0\,(y_i) = 0, \text{ for all } f_0 \in Ker\,(\Phi)\,,$$

then for $f_0 := (\Phi\,(f))\, f^* - (\Phi\,(f^*))\, f$,

$$\Phi\,(f) \sum_{i=1}^{r} \alpha_i\,(f^*\,(y_i))^2 = \Phi\,(f^*) \sum_{i=1}^{r} \alpha_i f^*\,(y_i)\, f\,(y_i)\,.$$

Now for $y_i \in \mathbf{Cr}\,(f)$

$$|\, f^*\,(y)| = \|f^*\|_{X,\infty}\ .$$

Also $\|f^*\|_{X,\infty} = 1$ and therefore, from $|f^*\,(y)| = \|f^*\|_{X,\infty}$, it follows that

$$f^*\,(y) = sign\,(f^*\,(y)) := \frac{f^*\,(y)}{|f^*\,(y)|}.$$

Therefore,

$$\Phi(f)\sum_{i=1}^{r}\alpha_i\left(f^*(y_i)\right)^2=\Phi(f^*)\sum_{i=1}^{r}\alpha_i sign\left(f^*(y_i)\right)f(y_i)$$

it implies that

$$\Phi(f)\sum_{i=1}^{r}\alpha_i=\Phi(f^*)\sum_{i=1}^{r}sign\left(f^*(y_i)\right)\alpha_i f(y_i).$$

Using $\Phi(f^*)=|||\Phi|||$, we obtain

$$\Phi(f)\sum_{i=1}^{r}\alpha_i=|||\Phi|||\sum_{i=1}^{r}sign\left(f^*(y_i)\right)\alpha_i f(y_i).$$

Let

$$a_i:=|||\Phi|||\frac{sign\left(f^*(y_i)\right)\alpha_i}{\sum_{i=1}^{r}\alpha_i}. \qquad [5.18]$$

We obtain that

$$\Phi(f)=\sum_{i=1}^{r}a_i f(y_i).$$

and

$$\sum_{i=1}^{r}|a_i|=|||\Phi|||\sum_{i=1}^{r}\frac{|\alpha_i|}{\left|\sum_{i=1}^{r}\alpha_i\right|}=|||\Phi|||.$$

as sought.

DEFINITION 5.3.– The expression $\Phi(f)=\sum_{i=1}^{r}a_i\, f(y_i)$, with $|||\Phi|||=\sum_{i=1}^{r}|a_i|$ and where the a_i's are defined in [5.18], is the canonical representation of the linear functional Φ.

PROPOSITION 5.3.– If f^* is extreme for Φ then for any canonical representation, we have

$$f^*(y_i)=sign(a_j),j=1,...,r.$$

PROOF.– From the above theorem (theorem 5.8)

$$\sum_{i=1}^{r} |a_i| = |||\Phi||| = \Phi(f^*) = \sum_{i=1}^{r} a_i f^*(y_i).$$

Hence

$$\sum_{i=1}^{r} |a_i| = \sum_{i=1}^{r} a_i f^*(y_i),$$

i.e.

$$\frac{|a_i|}{a_i} = f^*(y_i)$$

which implies

$$f^*(y_i) = sign(a_j).$$

Finally, we obtain

$$\|f^*\|_{C^{(0)}([a,b])} = 1. \qquad \square$$

5.7.3. *The specific case of the Dirac functional at point* 0

Determine the extreme points of the Dirac function δ_0,

$$\delta_0 : \left(\mathcal{P}_{g-1}(X), \|.\|_{X,\infty}\right) \to (\mathbb{R}, |.|);\, P \mapsto \delta_0(P) := P(0).$$

We make use of the canonical representation of δ_0. It is defined through g points $y_1 < ... < y_{g1}$ in X which are associated to coefficients $\alpha_1, ..., \alpha_g$ in $\mathbb{R}$, such that

$$\delta_0(P) = \sum_{j=1}^{g} \alpha_j P(y_j), \text{ for all } P \in \mathcal{P}_{g-1}(X)$$

and

$$|||\delta_0||| = \sum_{j=1}^{g} |\alpha_j|.$$

Consider the elementary Lagrange polynomials

$$l_i(y) := \prod_{j=1, i\neq j}^{g} \frac{y - y_j}{y_i - y_j}, i = 1, ..., n+1.$$

Then,

$$\delta_0(l_i) = l_i(0) = \prod_{j=1, i\neq j}^{g} \frac{-y_j}{y_i - y_j}.$$

Since,

$$l_i(y_j) = \begin{cases} 1 \text{ if } i = j \\ 0 \text{ if } i \neq j \end{cases}$$

it results that

$$\delta_0(l_i) = \sum_{i=1}^{g} \alpha_j l_i(y_j) = \alpha_i.$$

Therefore,

$$\alpha_i = \prod_{j=1,\, j\neq i}^{g} \frac{-y_j}{y_i - y_j}, \text{ for all } i = 1, ..., g.$$

We will need the following fact.

LEMMA 5.1.– Assume now that $0 \notin X$. In such case none of the α_j is null and

$$\alpha_i \alpha_{i+1} > 0 \text{ if and only if } y_i \leq 0 \leq y_{i+1}. \qquad [5.19]$$

PROOF.–

$$\alpha_i \alpha_{i+1} = \left(\prod_{j=1,\, j\neq i}^{g} \frac{-y_j}{y_i - y_j} \right) \left(\prod_{j=1,\, j\neq\, i+1}^{g} \frac{-y_j}{y_{i+1} - y_j} \right)$$

$$= \left(\frac{(-1)^{n+1} \prod_{j=1}^{g} y_j}{y_i} \right)^2 \frac{1}{\left(\prod_{j=1,\, j\neq i}^{g} (y_i - y_j) \right) \left(\prod_{j=1,\, j\neq i}^{g} (y_i - y_j) \right)}$$

The sign of this expression is determined by the product

$$\left(\prod_{j=1,\, j\neq i}^{g} (y_i - y_j)\right)\left(\prod_{j=1,\, j\neq i}^{g} (y_i - y_j)\right)$$

It follows that [5.19] holds. □

If P^* is an extreme point for δ_0, then

$$0 < |||\delta_0||| = \delta_0\left(P^*\right).$$

Furthermore, by the canonical representation

$$\delta_0\left(P^*\right) = \sum_{j=1}^{g} \alpha_j P^*\left(y_j\right).$$

Therefore,

$$0 < |||\delta_0||| = \left|\delta_0\left(P^*\right)\right| = \left|\sum_{j=1}^{g} \alpha_j P^*\left(y_j\right)\right|.$$

Since,

$$\alpha_i = \prod_{j=1,\, j\neq i}^{g} \frac{-y_j}{y_i - y_j} = l_i\left(0\right),\ \text{for all } i = 1, ..., n+1$$

and since $P^*\left(y\right)$ is determined through its values on g distinct points in X, we may obtain it through

$$P^*\left(y\right) = \sum_{i=1}^{g} l_i\left(y\right) P^*\left(y_i\right).$$

Now note that this latest expression characterizes P^* in a unique way. The polynomials $l_i\left(y\right)$ are a based on $\mathcal{P}_{g-1}$. Since the α_i are not identically 0 for all i, from lemma 5.1, it follows that all extreme polynomials P^*s coincide, from where it draws its uniqueness.

Assuming $0 < y_1$ or $0 > y_{n+1}$, the set of points $\{y_1, ..., y_g\}$ is an alternating set for P^*. This implies that P^* is a multiple of the Chebyshev polynomial in $X \cap [y_1; y_g]$.

PROPOSITION 5.4.– The solution of the minimax problem for the null function on X in $\mathcal{P}_{g-1}$ coincides with the unique extreme solution for the Dirac function δ_0.

This central result amounts to say that the optimal design has support defined by the unique extreme solution of δ_0. This support is the set of the oscillating points of this polynomial.

5.7.4. *Remez algorithm for the extreme polynomial: the optimal design in general cases*

Up to a translation we may choose $c = 0$, hence the case considered by de Boor and Rice. We thus have $b < 0$ and $d > 0$. With this choice, the algorithm which asymptotically produces the minimax polynomial and therefore, the optimal design, is written as:

1) choose $\mathbf{t} = (t_0, ..., t_g)$ in X, strictly increasing, and with $t_k = b$, $t_{k+1} = d$ for some k;

2) set $P := \sum_{i=0}^{g-1} (sign\,(l_i\,(0)))\, l_i$, with $l_i\,(x) := \prod\limits_{i \neq j} \frac{x - x_j}{x_i - x_j}$;

3) set $t_0 := \min X$, $t_{g+1} := \max X$ and construct s by

$$s_j := \begin{cases} t_j \text{ for } j = k, k+1 \\ \text{the first of the possibly two maxima of } P\,(t_j)\, P\text{in } [t_{j-1}, t_j] \cap X \end{cases}$$

for $j = 0, ..., k-1, k+2, ..., g$.

Choose $\widetilde{\mathbf{t}}$ from s as follows: a) if $P\,(t_0)\, P\,(t_1) < -1$, then $\widetilde{\mathbf{t}} := (t_0, s_1, ..., s_g)$, and increase k by 1, b) if $P\,(t_0)\, P\,(t_g) < -1$, then $\widetilde{\mathbf{t}} := (s_2, ..., s_g, t_{g+1})$, and decrease k by 1, c) else, $\widetilde{\mathbf{t}} := \mathrm{s}$.

5) Set $\mathbf{t} := \widetilde{\mathbf{t}}$.

6) Repeat steps 2 to 5.

The above properties of the extreme polynomials lead to the following result:

THEOREM 5.9.– The above algorithm converges to P^*.

PROOF.– See de Boor and Rice [BOO 82, pp. 51–52]. □

The equioscillation points of P^* determine the support of the optimal design.

The implementation of the algorithm, producing the extreme polynomial, can be performed using Sauer's code; see [SAU 83].

5.8. Spruill theorem: the optimal design

Turning back to the optimal design problem, and denoting here P the regression polynomial, Spruill [SPR 84b] obtains the following characterization of the optimal design.

THEOREM 5.10.– The optimal design for estimating $P(u)$ using the least squares estimator is supported at the proper oscillation points of $P^* \in \mathcal{P}_{g-1}(1)$, a set of either g or $g+1$ points, independent of c, and containing b and and at least one of a and e. If there are g points $x_1 < ... < x_g$, then the optimal proportions are

$$\frac{|l_j(c)|}{\sum_{j=0}^{g} |l_j(c)|}$$

where l_j are the Lagrange interpolation polynomials of degree $g-1$ to the points $\{x_1, ..., x_g\}$. If there are $g+1$ points then, every optimal design is a convex combination of the two g point designs formed as above from $\{x_1, ..., x_g\}$ and $\{x_2, ..., x_{g+1}\}$.

6

Instability of the Lagrange Interpolation Scheme With Respect to Measurement Errors

6.1. Introduction

For a fixed number of measurements n, all designs depend on g, the number of nodes x_is, and on n_i, which is the number of measurements to be performed at node x_i, under the constraint $n = n_0 + .. + n_{g-1}$. For unbiased designs, i.e. when the function f is a polynomial with degree $g-1$, the parameters $(x_i, n_i), i = 0, .., g-1$ define the variance of the resulting estimators. However, in the case when f is a straight line, all estimators based on Lagrange extrapolation or interpolation schemes propagate measurement errors in a way that may lead to strong instability (see section 3.8). The present chapter aims to analyze these effects and discuss possible solutions.

In the first section, we will consider the nature of uncertainty and inaccuracy, and the role of the distribution of the nodes on the stability of the design with respect to the propagation of those errors. The second section explains the definition of the total number n of experiments which provides a bound for the probability that the resulting estimators enjoy a prescribed degree of confidence.

6.2. The errors that cannot be avoided

These errors stem from two causes:

1) the numerical approximations;

2) the randomness of the data.

Consider a generic real number x with respect to the first point. We may write $x = s(x)\, mb^q$. The integer number b is called the base, 2 or 10, whereas m is the mantissa and $s(x) \in \{-1, 0, 1\}$ is the sign of x, and q is called the exponent. When $|m| \in \left]\frac{1}{b}, 1\right[$, then mb^q is called the normal representation of x.

A real number is written through the triple $(s(x), m, q)$. Since the number of digits is bounded in any computer, the collection of triples is bounded. Two types of triple are considered: simple and double precision.

The simple precision is defined by the fact that the representation $(s(x), m, q)$ requires only 32 bits, whereas the double precision requires 64 bits of information. It follows that the representation $\widetilde{x}$ for a given real number x is usually an approximation of x. For example, the real number π with representation with eight digits in base 10 admits the representation,

$$\pi \approx 0,31415927 \cdot 10^1.$$

We may quantify the error committed substituting x by $\widetilde{x}$, through $|\varepsilon|$, where

$$\varepsilon := \frac{x - \widetilde{x}}{x}.$$

Therefore,

$$\widetilde{x} = x(1 + \varepsilon). \qquad [6.1]$$

Note that this relation pertains to the relative error due to the rounding or truncating mechanism. Also, there exist two ways to approximate a number: the first is by truncating (we discard all figures in the mantissa following the first n ones), while the other one is by rounding (we add $1/2$ to the $n-$th figure in the mantissa). Obviously, these two modes of approximating a real number do not bear the same consequences. Consider, for example $x = mb^q$, and let n be the number of figures which is used in order to code m. In the case of truncation, we obtain $|\varepsilon| \leq b^{1-n}$. For the rounding approximation, we obtain $|\varepsilon| \leq \frac{1}{2} b^{1-n}$ which proves that rounding is a better procedure.

The number

$$eps := \frac{1}{2} b^{1-n},$$

is called the machine accuracy.

Clearly, when working with numbers, either rounded or truncated, the results of operations will also bear some errors. Any algorithm may increase the input errors.

To consider the propagation mechanism, denoting $ar\left(x\right)$ the rounded value of x, it holds, for elementary operations

$$\begin{aligned} ar\left(ar\left(x\right)+ar\left(y\right)\right) &= \left(ar\left(x\right)+ar\left(y\right)\right)\left(1+\varepsilon_1\right), \\ ar\left(ar\left(x\right)-ar\left(y\right)\right) &= \left(ar\left(x\right)-ar\left(y\right)\right)\left(1+\varepsilon_2\right) \\ ar\left(\frac{ar\left(x\right)}{ar\left(y\right)}\right) &= ar\left(1+\varepsilon_3\right), \\ ar\left(ar\left(x\right)ar\left(y\right)\right) &= ar\left(x\right)ar\left(y\right)\left(1+\varepsilon_4\right), \end{aligned}$$

where,

$$|\varepsilon_i| < eps, i = 1, 2, 3, 4.$$

Rounding may lead to some increase of the error by simple operations. For example, evaluate

$$\sqrt{x+\frac{1}{10^9}}-\sqrt{x}.$$

By rounding we get 0.

However, we may avoid this error noting that

$$\left(\frac{\sqrt{x+\frac{1}{10^9}}-\sqrt{x}}{\sqrt{x+\frac{1}{10^9}}+\sqrt{x}}\right)\left(\sqrt{x+\frac{1}{10^9}}-\sqrt{x}\right)=\frac{\frac{1}{10^9}}{\sqrt{x+\frac{1}{10^9}}+\sqrt{x}}.$$

We now provide a definition for the stability of an algorithm, which is considered as a function of the input data. Let x be input data and $\widetilde{x}$ be its approximation as used for the calculation of the function f. The sensitivity of f with respect to this approximation is measured through

$$\frac{\frac{f(x)-f(\widetilde{x})}{f(x)}}{\frac{x-\widetilde{x}}{x}}$$

for small $x-\widetilde{x}$.

When f is a regular function, then

$$\left| \frac{\frac{f(x)-f(\widetilde{x})}{f(x)}}{\frac{x-\widetilde{x}}{x}} \right| \sim \left| \frac{xf'(x)}{f(x)} \right| .$$

This yields the following definition.

DEFINITION 6.1.– Assume that f belongs to $\mathcal{C}^{(1)}([a,b])$. For x in (a,b), the number

$$\kappa_x(f) := \left| \frac{xf'(x)}{f(x)} \right|$$

is the condition number of the function f at point x.

EXAMPLE 6.1.– Let $f(x) = \sqrt{x}$. Then $\kappa_x(f) := \left| \frac{xf'(x)}{f(x)} \right| = \frac{1}{2}$. The function f admits a good condition number as the relative error on $f(x)$ is half the relative error on x. In the case when $f(x) = k - x$, then $\kappa_x(f) = \left| \frac{x}{x-k} \right|$, which yields a bad condition number when x is close to k.

The function f may have a good condition number, but the algorithm which evaluates $f(x)$ may include operations which, in turn, have bad condition numbers.

EXAMPLE 6.2.– Let $f(x) = \sqrt{x+1} - \sqrt{x}$. Then $\kappa_x(f) := \left| \frac{xf'(x)}{f(x)} \right| = \frac{1}{2}\sqrt{\frac{x}{x+1}}$. When x is "large", $\kappa_x(f) \sim \frac{1}{2}$. Consider the evaluation of $f(12345)$ in a straight way. We obtain $f(12345) = \sqrt{12346} - \sqrt{12345} \sim \frac{10^{-2}}{2}$. A sharp calculus yields $f(12345) = 0,4500032\ 10^{-2}$. Henceforth, the relative error is of order 10%. We may also use the algorithm $f(x) = \frac{1}{\sqrt{x+1}+\sqrt{x}}$ and a six digit representation for the argument. This provides $f(12345) = 0,450002 \cdot 10^{-2}$. The relative error has been decreased to 0.0003%.

6.2.1. *The role of the errors: interpolation designs with minimal propagation of the errors*

We now consider the role of the above errors due to rounding or truncating the optimal design for interpolation, neglecting the random term in the model, which is therefore written as,

$$y_i = f(x_i)$$

where x_i denotes a node and f is the unknown function to be estimated. The measurements of the function f at points x_is are known up to rounding or truncating effect, introducing, therefore, a term ϵ_i so that the value $f(x_i)$ is substituted by $\widetilde{f(x_i)} = f(x_i) + \epsilon_i$. Since no random effect alters the quality of the measurement $f(x_i)$ it is assumed that a single observation is performed on the node x_i.

We assume that the values of the nodes x_i are known without any type of error, neither rounding nor truncating. The numerical errors only stem from the algorithm which provides the values of the function f on the nodes. This approach is, indeed, the one which is adopted in physics or in engineering for fixed designs commonly: the measurements are fixed, with no uncertainty. No randomness is assumed on the measurement points x_i, in contrast to the case of random design experiments.

If we were to consider sharp uncertainties pertaining to the values of the nodes, we would use the tools indicated in Chapter 3; we would then attribute this to the instability of the interpolation scheme, using the fact that the Lagrange polynomials is in direct relation with the determinant of the Vandermonde matrix (see section 3.1)

$$G = \begin{pmatrix} x_0^0 & . & x_0^j & . & x_0^{g-1} \\ . & . & . & . & . \\ x_i^0 & . & x_i^j & . & x_i^{g-1} \\ . & . & . & . & . \\ x_{g-1}^0 & . & x_{g-1}^j & . & x_{g-1}^{g-1} \end{pmatrix}$$

and implies $(2g+1)(g+1)$ operations. This would yield to other interpolation schemes, using for example, Newton polynomials; see section 3.8.

To quantify the propagation of the error on the values of the $f(x_i)$, assume that we work with simple precision. The order of magnitude of the relative error is

$$eps = 5,96 \cdot 10^{-8}.$$

Let $\{x_i : i = 0, ..., g-1, g \in \mathbb{N}^*\}$ be the nodes in $[a,b]$. The stability of the Lagrange interpolation scheme is stated as follows:

DEFINITION 6.2.– Let $X := \{x_i : i = 0, ..., g-1, g \in \mathbb{N}^*\}$ be the nodes in $[a,b]$. Let f be a continuous function defined on $[a,b]$. The Lagrange interpolation scheme is stable when the Lebesgue constant

$$\Lambda_{g-1}(X) := \max_{x \in [a,b]} \sum_{i=0}^{g-1} |l_i(x)| \leq 1.$$

We will then find that the Lagrange interpolation scheme cannot be proved to be stable according to the above definition 6.2. The only available upper bound for

$$\Delta_g = \left| \sum_{i=0}^{g-1} \widetilde{f(x_i)} l_i(x) - \sum_{i=0}^{g-1} f(x_i) l_i(x) \right|$$

will tend to infinity as the number of nodes g increases. We prove that the Lebesgue constant is the condition number of the Lagrange scheme.

We evaluate an upper bound for the errors on the output Δ_g. It holds

$$\begin{aligned} \Delta_g &= \left| \sum_{i=0}^{g-1} \left(f(x_i)(1+\varepsilon_i) - f(x_i) \right) l_i(x) \right| = \left| \sum_{i=0}^{l} f(x_i) \varepsilon_i l_i(x) \right| \\ &\leq \sum_{i=0}^{l} |f(x_i) \varepsilon_i l_i(x)| \\ &\leq \left(\max_{i=0,\ldots,g-1} |f(x_i)| \right) \Lambda_g(X) \max_{i=0,..,g-1} \left| \widetilde{f(x_i)} - f(x_i) \right| . \end{aligned} \quad [6.2]$$

where

$$\Lambda_g(X) := \left(\max_{x \in [a,b]} \sum_{i=0}^{g-1} |l_i(x)| \right) .$$

The term $\max_{i=0,..,g-1} \left| \widetilde{f(x_i)} - f(x_i) \right|$ is the maximal input error.

The bound in [6.2] diverges to infinity as g increases, but in the case when the relative error $|\varepsilon_i|$ equals zero. Minimizing the upper bound with respect to the nodes leads to the Chebyshev nodes; see section 3.8 and the arguments hereafter.

The only possibility to minimize the maximal risk due to the lack of accuracy in the collect of the data $f(x_i)$ as such is to choose nodes which do not lead to a fast divergence of the bound as g increases. This fact justifies the choice of the Chebyshev nodes to reduce instability. We present two examples in this respect; the first one shows that the Lagrange interpolation scheme may be instable; in the second example we show the the instability depends on the location of the nodes.

EXAMPLE 6.3.– We recall the example 3.3. We interpolate the function $f(x) = \sin(2\pi x)$ making use of 22 nodes, equally sparsed on $[-1, 1]$. We perturb the function f on each of these nodes, leading to $\widetilde{f}$ such that

$$\max_{i=0,..,21} \left| f(x_i) - \widetilde{f}(x_i) \right| \backsimeq 9.510^{-4}.$$

The direct evaluation is as follows:

$$\max_{x\in[-1,1]} \left| \mathcal{L}(f)(x) - \mathcal{L}\left(\widetilde{f}\right)(x) \right| \backsimeq 3.1342.$$

Clearly, the resulting error far exceeds the input one, proving instability; see, Chapter 7 in [QUA 07].

EXAMPLE 6.4.– For $g = 32$ and $f(x) = \sin x$ on $[0, 5]$ it holds: the input data are obtained in simple precision; the output ones in double precision; with a double precision calculation, the Lebesgue constant Λ_g satisfies

$$\Lambda_{32}(X_1) := \max_{x\in[0,5]} \sum_{i=0}^{31} |l_i(x)| \geq 1, X_1 := \left\{ x_i := a + \frac{i}{g}(b-a) \right\}$$

$$\text{and } \Lambda_g(X_1) \sim \frac{2^{g+1}}{eg(\log g + \gamma)}$$

for equally sparsed nodes, proving instability;

$$\Lambda_{32}(X_2) := \max_{x\in[0,5]} \sum_{i=0}^{31} |l_i(x)| \leq 1, X_2$$

$$:= \left\{ x_i := \frac{a+b}{2} + \frac{b-a}{2} \cos\left(\frac{2i+1}{2g-1}\pi \right) \right\}$$

and $\Lambda_g(X_2) \sim \frac{2}{g} \ln g$ for the Chebyshev nodes, proving stability.

Let $f(x_i)$ be the exact value of the function f at the $i-th$ node. The calculation was done using $\widetilde{f(x_i)} = f(x_i)(1+\varepsilon_i), |\varepsilon_i| < 5,96 \cdot 10^{-8}$.

6.2.2. *Optimizing on the nodes*

We now consider the optimal nodes with respect to the upper bound [6.2].

Until now we have considered two classes of node: the equidistributed ones and the roots of the Chebyshev polynomials of first kind. For a given number of nodes a better choice can be proposed, i.e. the extended Chebyshev nodes, which are defined on $[-1,1]$, by

$$x_i := \frac{\cos\frac{(2i-1)\pi}{2(g-1)}}{\cos\frac{\pi}{2(g-1)}}, i = 1,...,g-1; g = 2,3,4,....$$

Let us consider that $\breve{T}$ is the set of the extended Chebyshev nodes and $\widetilde{T}$ the Chebyshev nodes in $[-1,1]$.

Let

$$l_i(x) := \prod_{i,j\neq i}^{g-1} \frac{x-x_i}{x_i-x_j}.$$

Brutman [BRU 97] proved that for $g \geq 2$

$$\max_{x\in[-1,1]} \sum_{i=0}^{g-1} |l_i(x)| < 0,7219 + \frac{2}{\pi}\ln g, \text{ for any } g \in \mathbb{N}^*.$$

The proof of the above result lies in some elementary properties of the function.

$$\lambda_g(X,x) := \sum_{i=0}^{g-1} |l_i(x)|,$$

where, $X := \{x_i \in [-1,1] : i = 0,...,g-1\}$ is a set of nodes. The function $\lambda_g(X,x)$ is called the *Lebesgue function*, whose maximal value upon x is $\Lambda_g(X)$.

The main properties of $\lambda_g(X,x)$ are as follows:

– for $g > 3$ the function $\lambda_g(X,x)$ is a piecewise polynomial satisfying $\lambda_n(X,x) \geq 1$ with equality only at the nodes x_i;

– $\lambda_g(X,x)$ has one local maximum on some which will be denoted by $\mu_k(X)$;

– $\lambda_g(X,x)$ is monotone decreasing and convex in $(-1,x_0)$ and monotone increasing and convex in $(x_{g-1},1)$. If $-1 < x_0$ and $x_{g-1} < 1$, we will denote $\lambda_g(X,-1)$ and $\lambda_g(X,1)$ by $\mu_0(X)$ and $\mu_g(X)$ correspondingly. The largest and the smallest local maxima of $\lambda_g(X,x)$ in $[-1,1]$ will be denoted correspondingly by $M_g(X)$ and $m_g(X)$.

Let $\delta_g(X) = M_g(X) - m_g(X)$.

– Let $Z = \{z_i : i = 0, ..., g-1\}$ and $Y = \{y_i : i = 0, ..., g-1\}$ be two sets of interpolations nodes related by an affine transformation $y_i = \alpha z_i + \beta$. Then, $\lambda_g(Z, z) = \lambda_g(Y, \alpha z + \beta)$.

Let $\gamma = 0,577...$ be the Euler constant. Then, as $g \to \infty$, it holds

$$\Lambda_g\left(\widetilde{T}\right) \sim \frac{2}{\pi} \log g \text{ (Bernstein [BER 18])}$$

$$\Lambda_g\left(\widetilde{T}\right) = \frac{2}{\pi} \log(g-1) + \frac{2}{\pi}\left(\gamma + \log\frac{8}{\pi}\right) + \mathcal{O}\left(\frac{1}{g^2}\right) \text{ (Brutman [BRU 97])}$$

For any fixed $g \geq 2$, it holds

$$\begin{aligned}\Lambda_g\left(\widetilde{T}\right) &= \lambda_g\left(\widetilde{T}, 1\right) = \lambda_g\left(\widetilde{T}, -1\right) \\ &= \frac{1}{g-1} \sum_{i=1}^{g-1} \cot \frac{2i-1}{4(g-1)} \pi \text{ (Zeller and Ehlich [EHL 65]).}\end{aligned}$$

We refer to Luttmann and Rivlin [LUT 65] for the above results.

We consider the choice of optimal nodes for the interpolation problem under errors on the input values $f(x_i)$.

From Chapter 3 (see section 3.5) it holds

$$\|f - \mathcal{L}(f)\|_\infty \leq (1 + \Lambda_g(X)) \min_{P_{g-1} \in \mathcal{P}_{g-1}} \|f - P_{g-1}\|_\infty .$$

Define by $\mathcal{L}(f)$ the Lagrange interpolation polynomial evaluated on the nodes defined as the set X in $[-1, 1]$, X, which contains g distinct points.

To reduce the approximation error the optimal nodes solve

$$\begin{aligned}X^* &= \arg\min_X \left((1 + \Lambda_g(X)) \min_{P_{g-1} \in \mathcal{P}_{g-1}} \|f - P_{g-1}\|_\infty\right) \\ &= \arg\min_X (1 + \Lambda_g(X)),\end{aligned}$$

where X is a finite subset of g distinct points in $[-1, 1]$. A general interval $[a, b]$ can be considered through an affine transformation.

$\Lambda_g^* := \Lambda_g(X^*)$ is the minimum value of $\Lambda_g(X)$ when X runs in all collections of g nodes in $[-1,1]$.

For fixed X with $cardX = g$, the Lebesgue function $\lambda_g(X,x)$ admits a local maximum between two consecutive nodes; see [BRU 97]. In the open interval $(x_{k-1,}x_k)$, $1 \leq k \leq g$, denote $\mu_k(X)$ this local maximum.

A result by Erdös [ERD 58b] states that

$$\frac{2}{\pi}\log g - c_1 \leq \Lambda_g^* \leq \frac{2}{\pi}\log g + c_2,\ g = 1, 2, \ldots.$$

Vértesi [VÉR 90] and Erdös and Vertesi [ERD 80] and [ERD 81] proved for $g \rightarrow \infty$,

$$\Lambda_g^* = \frac{2}{\pi}\log(g-1) + \frac{2}{\pi}\left(\gamma + \log\frac{4}{\pi}\right) + \mathcal{O}\left(\left(\frac{\log\log g}{\log g}\right)^2\right).$$

The characterization of the set of nodes X^* has been handled by Bernstein [BER 18] and by Erdös [ERD 58b]. They stated the following conjectures, which compares the performances of system of nodes with different cardinality.

– Bernstein conjecture:

$$X^* \text{ is such that } \mu_i(X^*) = \mu_1(X^*),\ i = 1, \ldots, g.$$

– Erdös conjecture, which strengthens Bernstein's conjecture:

$$\min_{i=1,\ldots,g} \mu_i(X) \leq \Lambda_g^* :\leq \max_{i=1,\ldots,g} \mu_i(X),$$

for all X with $cardX = g$.

The proofs of those conjectures have been obtained by Kilgore, de Boor and Pinkus (see [BRU 97] for details).

Brutman [BRU 97] proved that for any $g \geq 2$

$$\frac{1}{2} + \frac{2}{\pi}\log g < m_g\left(\widetilde{T}\right) < \Lambda_g^* < M_g\left(\widetilde{T}\right) < \frac{3}{4} + \frac{2}{\pi}\log g.$$

Finally, the Erdös theorem, (see Theorem 3.11 in Chapter 3) states that asymptotically X^* coincides with $\widetilde{T}$ which provides a further strong argument in favor of the Chebyshev nodes.

REMARK 6.1.– There is no way to obtain some stability with respect to the accuracy of the calculation of the output $f(x_i)$ on a node. Uncertainty due to randomness, on the contrary, may be controlled and countered in a number of ways: 1) choosing an optimal estimator in a given class; 2) choosing optimal nodes with respect to the divergence of the Lebesgue constant; 3) reducing the variability of the data, increasing the number of replications of the measurements; 4) changing the estimation scheme. For this last sake, a spline scheme may be preferred, adapting polynomials with low degree on any subinterval of the interpolation domain, and respecting the smoothness boundary conditions. Now, extrapolation is known to be well defined only for analytic functions (see Chapter 3), a class of functions which can be accurately approximated by polynomials with low degree. Also, the inequality of Coatmélec implies that the Lagrange interpolation polynomial of the derivative of any order of any analytic function converges together the degree of the polynomial, to the derivative (with same order); the convergence is quite fast, even using Lagrange polynomials with low degree for the approximation; see Chapter 3 [3.9]. The approximation can be obtained using an observation domain with small length, and provides a global approximating function. In this context, the Lagrange scheme also provides a consistent tool for the extrapolation, without requirement of smoothness conditions as for the splines with respect to the behavior of the approximation at the boundaries of the grid.

6.3. Control of the relative errors

For a continuous function f on the interval $[a, b]$, assuming that the f does not take value 0 on $[a, b]$, taking P^{**} the best uniform approximating polynomial of f among all polynomials in $\mathcal{P}_n$, where the optimality holds with respect to the relative error, viz.

$$P^{**} := \arg \inf_{P \in \mathcal{P}_n} \left\| \frac{f - P}{f} \right\|_\infty .$$

In this section, we prove the existence and uniqueness of P^{**} and we show that whenever for some polynomial P in $\mathcal{P}_n$

$$\frac{f(x_i) - P(x_i)}{f(x_i)} = (-1)^i E^{**}$$

on a family of pairwise distinct points x_i, $i = 0, ..., n+1$, on $[a, b]$, then $P = P^{**}$ and

$$E^{**} = \inf_{P \in \mathcal{P}_n} \left\| \frac{f - P}{f} \right\|_\infty ,$$

namely, the infimum relative error of the approximation scheme on $[a, b]$. Consider

$$R^{**}(x) := \left(\frac{f-P}{f}\right)(x)$$

as the relative error function.

Assume that there exists some $Q \in \mathcal{P}_n$ with $e := \left\|\frac{f-Q}{f}\right\|_\infty < \left\|\frac{f-P^{**}}{f}\right\|_\infty$.

Let $S := P^{**} - Q$; then write

$$\begin{aligned}
S(x_i) &= P^{**}(x_i) - Q(x_i) = f(x_i) - f(x_i) + P^{**}(x_i) - Q(x_i) \\
&= (f(x_i) - Q(x_i)) + (-f(x_i) + P^{**}(x_i)) \\
&= (f(x_i) - Q(x_i)) - (f(x_i) - P^{**}(x_i)) \\
&= \left[\left(\frac{f(x_i) - Q(x_i)}{f(x_i)}\right) - \left(\frac{f(x_i) - P^{**}(x_i)}{f(x_i)}\right)\right] f(x_i) \\
&= \left(\frac{f(x_i) - Q(x_i)}{f(x_i)} - \left((-1)^i E^{**}\right)\right) f(x_i) \\
&= \left(\frac{f(x_i) - Q(x_i)}{f(x_i)} + (-1)^{i+1} E^{**}\right) f(x_i).
\end{aligned}$$

For even values of i

$$S(x_i) = \left(\frac{f(x_i) - Q(x_i)}{f(x_i)} - E^{**}\right) f(x_i) = (e - E^{**}) f(x_i)$$

with $e - E^{**} < 0$. When i is odd then,

$$S(x_i) = \left(\frac{f(x_i) - Q(x_i)}{f(x_i)} + E^{**}\right) f(x_i) = (e + E^{**}) f(x_i)$$

with $e + E^{**} > 0$. Hence, the polynomial S changes sign $n + 2$ times as i increases to $i + 1$ (it goes from $-f(x_i)$ to $f(x_i)$), hence S has at least $n + 1$ roots. But deg $(S) \leq n$, since $S \in \mathcal{P}_n$ and therefore $S(x) = 0$, for all $x \in [a, b]$. Hence, $Q = P^{**}$ and thus $\left\|\frac{f-Q}{f}\right\|_\infty < \left\|\frac{f-P^{**}}{f}\right\|_\infty$ which cannot hold.

The proof of the reciprocal claim follows the same line as given in the proof of the classical Borel–Chebyshev theorem.

EXAMPLE 6.5.– Let $f \neq 0$ in $[a, b]$, and find the best relative approximation among all polynomials with degree less than or equal to 1. With $P^{**}(x) = \alpha + \beta x$, there should exist three points, x_0, x_1, x_2, of maximal and minimal values of

$$R^{**}(x) = \frac{f(x) - \alpha - \beta x}{f(x)}.$$

One of those is necessarily an interior point

$$\left(\frac{dR^{**}(x)}{dx}\right)_{x=x_1} = 0$$

and thus

$$\frac{f(x_1)}{\left(\frac{df(x)}{dx}\right)_{x=x_1}} - x_1 = \frac{\alpha}{\beta}.$$

Since a and b are equioscillation point, it holds

$$\begin{cases} \frac{f(a)-\alpha-\beta a}{f(a)} = E^{**} \\ \frac{f(x_1)-\alpha-\beta x_1}{f(x_1)} = -E^{**} \\ \frac{f(b)-\alpha-\beta b}{f(b)} = E^{**} \end{cases}.$$

Hence, x_1, α, β, E^{**} solve

$$\begin{cases} \frac{f(a)-\alpha-\beta a}{f(a)} = E^{**} \\ \frac{f(x_1)-\alpha-\beta x_1}{f(x_1)} = -E^{**} \\ \frac{f(b)-\alpha-\beta b}{f(b)} = E^{**} \\ \frac{\alpha}{\beta} = \frac{f(x_1)}{\left(\frac{df(x)}{dx}\right)_{x=x_1}} - x_1 \end{cases}.$$

We have

$$\begin{cases} f(a) - \alpha - \beta a = E^{**} f(a) \\ f(x_1) - \alpha - \beta x_1 = -E^{**} f(x_1) \\ f(b) - \alpha - \beta b = E^{**} f(b) \\ \frac{\alpha}{\beta} = \frac{f(x_1)}{\left(\frac{df(x)}{dx}\right)_{x=x_1}} - x_1 \end{cases}$$

If for example $f : [0,1] \to \mathbb{R}, x \mapsto f(x) := e^x$, then

$$\begin{cases} \alpha = 1 - E^{**} \\ e^{x_1} - \alpha - \beta x_1 = -E^{**} e^{x_1} \\ e - \alpha - \beta = E^{**} e \\ \frac{\alpha}{\beta} = \frac{f(x_1)}{e^{x_1}} - x_1 \end{cases}$$

Now what happens if we look for a polynomial with degree $n > 1$?

The nodes of interpolation, i.e. the points of equioscillation, may be obtained through a variation of the Remez algorithm.

At the **k**-th step the coefficients of the polynomial $P_n^{(\mathbf{k})}$ were defined, in the classical case (minimal absolute error of the approximation of the null function, see Chapter 1 [1.2]), by the system

$$f(x_i) - P_n^{(\mathbf{k})}(x_i) = (-1)^i E^{(\mathbf{k})}$$

as $i = 0, ..., n+1$.

In the present setting, the rest function is no longer

$$f - P_n^{(\mathbf{k})}$$

but merely

$$\frac{f(x_i) - P_n^{(\mathbf{k})}(x_i)}{f(x_i)}$$

and the equioscillation condition is written for $i = 0, ..., n+1$,

$$\frac{f(x_i) - P_n^{(\mathbf{k})}(x_i)}{f(x_i)} = (-1)^i E^{(\mathbf{k})}$$

i.e.

$$f(x_i) - P_n^{(\mathbf{k})}\left(x_i^{(\mathbf{k})}\right) = (-1)^i E^{(\mathbf{k})} f\left(x_i^{(\mathbf{k})}\right)$$

i.e.

$$f(x_i) = \sum_{j=0}^{n} a_j^{(\mathbf{k})} \left(x_i^{(\mathbf{k})}\right)^j + (-1)^i E^{(\mathbf{k})} f\left(x_i^{(\mathbf{k})}\right)$$

with

$$P_n^{(\mathbf{k})}(x_i) := \sum_{j=0}^{n} a_j^{(\mathbf{k})} \left(x_i^{(\mathbf{k})}\right)^j.$$

The minimum/maximum points $y_i(\mathbf{k})$ are relative extrema of the function

$$R_{\mathbf{k}}(x) := \frac{f(x) - P_n^{(\mathbf{k})}(x)}{f(x)}.$$

Under these conditions, convergence of the modified Remez algorithm holds; see section 2.5.

We propose an example for the calculation of the relative error in the Lagrange interpolation, using the modified Remez algorithm.

EXAMPLE 6.6.– Let $f : [0,1] \to \mathbb{R}, x \mapsto f(x) := e^x$. Determine $P_3^{**} \in \mathcal{P}_3$ minimax in the relative sense for e^x. Begin with the five Chebyshev points in $[0,1]$,

$$x_0^{(1)} = 0, x_1^{(1)} \sim \frac{14}{100}, x_2^{(1)} = \frac{1}{2}, x_3^{(1)} \sim \frac{85}{100}, x_4^{(1)} = 1.$$

From the system

$$f(x_i) = \sum_{j=0}^{3} a_j^{(1)} \left(x_i^{(1)}\right)^j + (-1)^i E^{(1)} f\left(x_i^{(\mathbf{k})}\right), i = 0,1,2,3$$

we obtain

$$a_0^{(1)} \sim \frac{99}{100}, a_1^{(1)} \sim \frac{101}{100}, a_2^{(1)} \sim \frac{43}{100}, a_3^{(1)} \sim \frac{27}{100}, E^{(1)} \sim \frac{3}{1000}.$$

The local extrema of $R_{\mathbf{1}}$ interior to $[0,1]$ are such that

$$\frac{dR_{\mathbf{1}}(x)}{dx} = 0$$

i.e.

$$\frac{d}{dx}\left(\frac{e^x - \left(0,99 + 1,01x + 0,43x^2 + 0,27x^3\right)}{e^x}\right) = 0$$

therefore

$$y_1 \sim \frac{13}{100}, y_2 \sim \frac{47}{100}, y_3 \sim \frac{83}{100}.$$

Hence for $\mathbf{k} = 2$ we consider the points

$$x_0^{(2)} = 0, x_1^{(2)} = y_1 \sim \frac{13}{100}, x_2^{(2)} = y_2 \sim \frac{47}{100}, x_3^{(2)} = y_3 \sim \frac{83}{100}, x_4^{(2)} = 1.$$

That the frontier points $\{0, 1\}$ are kept results from the fact that

$$\frac{d^4}{dx^4}\left(e^x - P_3^{**}(x)\right) = e^x \neq 0$$

on $[0, 1]$. With the usual stopping rule (modified taking into account the relative error), for $\mathbf{k} = 4$ the coefficients of $P_3^{(\mathbf{4})}$ are those of $P_3^{(3)}$ up to the third decimal. Hence,

$$P_3^{**}(x) = P_3^{(3)}(x) \sim 1 + x + 0,43x^2 + 0,27x^3.$$

6.3.1. *Implementation of the Remez algorithm for the relative errors*

We need to

1) fix "arbitrarily" $n+2$ pairwise distinct points $x_0, ..., x_{n+1}$ in $[a, b]$. It is a good start to begin with the Chebyshev nodes

$$x_i := \frac{a+b}{2} + \frac{b-a}{2}\cos\left(\frac{i\pi}{n+1}\right)$$

in $[a, b]$. This entails a quadratic rate of convergence (see Theorem 2.6);

2) solve the linear system with $n+2$ equations and $n+2$ unknown $a_0, ..., a_n, E$,

$$\begin{cases} f(x_0) - (a_0 + a_1x_0 + ... + a_nx_0^n) = E \\ \dots\dots\dots\dots\dots\dots\dots\dots\dots\dots \\ f(x_0) - (a_0 + a_1x_0 + ... + a_nx_0^n) = (-1)^{n+1} E \end{cases}.$$

This yields a unique polynomial P with degree n. If the degree of P is less than n, change the initial points and start again (see section 2.5.1);

3) determine the abscissae y_i of local extreme values of the function $(f - P)/f$ and start back with step 1 substituting points x_is by the points ys through the scheme described in Chapter 2.

The following is a Maple program due to Zimmerman–Lefévre (see [LEF 04])

```
remez := proc(f, x, n, a, b) local pts, i, sys, eps, p, c, q, oldq;

p := add(c[i] * x^i, i = 0..n);

pts := {seq(evalf(((a + b)/2 + (b - a)/2 * cos(Pi * i/(n + 1))), i = 0..n + 1)};
```

```
while q< > oldq do

sys := {seq(evalf(subs(x = op(i+1,pts), p - f)) = (-1)^i * eps, i = 0..n+1)};

sys := solve(sys, {seq(c[i], i = 0..n), eps});

oldq := q;

q := subs(sys, p);

lprint(q);

pts := sort([a, solve(diff(q - f, x), x), b]);

od;

q

end:
```

Another program is due to Muller; see Muller (pag 42,43)

```
Remez: =proc(f, x, n, a, b)

P: =add(p [i] * x^i, i = 0, ..., n);

pts: =sort([seq(evalf(1/2*a + 1/2b + 1/2(b - a)* cos(Pi * i/(n + 1))),
i = 0...n + 2)]);

#points

ratio: =2:

Count: =1:   threshold: =1.000005

while ratio > threshold do

sys: ={seq(oval f (subs(x = op(i + 1, pts), P - f))= (-1)^i * eps, i = 0...n + 1}

print f ("ITERATION NUMBER: %a\n*,COUNT);

print f ("Current list of points : %a\n*,pts);

Count: =Count+1;
```

```
print f (*Linear system : %a\n*sys );
sys: =solve(sys, {eps, seq (p [i] , i = 0...n)});
# we compute the polynomial associated with the list of
# points
  old q: =q;
        q: =subs{sys, P};
        print("Current list of points :%a\n*, q);
# we now compute the new list of points
# by looking for the extremes of q − f
derivative: =unapply(diff (q − f, x));
pts: =AllRootsOf(derivative, a,b);
no: =nops(pts);
if no > n + 2 then print ("Too many extreme values, tryl arger degree")
elif no = n then pts := [a, op (pts) , b]
elif no = n + 1 then
if abs ((q − f) (a)) >abs ((q − f) (b))
       then pts: = [a, op (pts)]
  else pts: = [op (pts) , b]
   end if
  elif no < n then print ("Not enough oscillations")
 end if;
lprint(pts);
```

```
E_max :=evalf (subs (x = pts [1] , abs (q − f)));
  E_min := E_max;
for i from 2 to (n + 2) do
Ecurr: =evalf (subs (x = pts [i] , abs (q − f)));
if Ecurr > E_max then E_max := Ecurr
elif Ecurr < E_min then E_min := E_curr fi
 end do;
ratio:= E_max/ E_min;
# We consider that we have found the Minimax polynomial
# ( i.e., that the conditions of Chebyshev's
# theorem are met)
# when 1 < E_max/ E_min threshold
# threshold must be very slightly above 1
print f ("error: %a\n", E_max);
 end do;
q
end proc.
```

Box 6.1. *Zimmerman Program*

Müller [MÜL 97, pp. 41–44] evaluates the relative minimax polynomial of degree 4 for $f : [0, 2] \to \mathbb{R}, x \mapsto \sin e^x$. The starting six points are

$$0, 0.1909830057, 0.6909830062, 1.309016994, 1.809016994, 2$$

which are the Chebyshev points

$$1 + \cos \frac{i\pi}{5}, i = 0, ..., 5$$

where it holds

$$|T_5(x-1)| = 1.$$

The minimax polynomial is found to be

$$P^*(x) = 0,6751752198 + 2,123585326x - 1,548341910x^2$$
$$- 2,293483579x^3 + 1,292440070x^4$$

and the relative error is

$$E := \max_{x\in[0,2]} \frac{|\sin e^x - P^*(x)|}{e^x} \leq 0,034.$$

6.4. Randomness

We now turn to the case when randomness comes into picture. Any measurement at node x_i is repeated n_i times. In interpolation the optimal design is well defined; it is the Chebyshev one, which achieves the double optimality requirement, namely on the variance of the estimator of the interpolated value, and on the Lebesgue constant, which controls the stability of the scheme.

We do not consider the stability of the algorithm as treated above any longer, but some control on the optimal design with respect to the features of the law of the errors. For this reason, we introduce some tools from the theory of the polynomial of a single variable, with some control on the maximal values of their derivatives on some interval. The main result of the next section, which is a theorem given by Markov, is the ingredient for the control of the number of replications which provide a given probability for the interpolated value to lay in a prescribed interval. Apart from its usefulness in the present context, the derivation of this result is of interest on its own.

From the discussion in the previous section we see irreverently of the fact that errors stem from deterministic effects of rounding or truncating or from random effects on the measurements, stability considerations yield to considerate a choice of the nodes which optimize the Lebesgue constant in relation to its increase (with respect to the distribution of the nodes). The Chebyshev family of nodes optimizes the variance of the interpolation and provides a minimal increase for the Lebesgue constant. However, the Lebesgue constant, which is an ingredient in the upper bound of the propagated error, is increasing with respect to the number of nodes. This number of replications has been considered a constraint of the design until now; here it is optimized in terms of a risk. It is, therefore, relevant to obtain the minimal number of nodes which are required to get a prescribed accuracy on the interpolation value, at least with a prescribed risk. This is the aim of this section.

We now consider the problem of the optimal value of n (which is the total number of observations), for which the error for extrapolated value does not exceed a prescribed value with a given probability.

Since the distribution of the data is unknown, we use some inequalities which we will state in the following section. The notation for the estimator and the design are as given in the previous chapters.

6.5. Some inequalities for the derivatives of polynomials

As for the notation, in this section we will assume that all roots of all trigonometric polynomials involved are defined modulo 2π, such that a root is the class of all of them, which differ from a multiple of 2π. For the sake of clarity of notation, the degree of the various polynomials involved in this section is indicated irreverently with respect to the notation pertaining to the designs; hence this section is purely instrumental; adapting those results to the context of the designs will be performed without difficulty when needed.

The coming discussion makes use of a basic theorem given by Markov, which provides information on the derivatives of a given polynomial when its upper bound on a compact interval is known; this result, associated with its complement by Bernstein, will be used to specify the number of experiments to be performed for a given accuracy of the resulting estimator. We state both results, whose proofs are in Appendix 3; the proofs, although technical, are of interest in their own, and owe much to deep work by Vitali. The change from a polynomial in the canonical basis onto a trigonometric polynomial through the change of variable $x = \cos\theta$ is the basic ingredient of this study, due to the controls of the trigonometric functions.

It holds

THEOREM 6.1 (Markov).– Let $P(x) := \sum_{j=0}^{n} a_j x^j$ be a polynomial with real coefficients and degree n. 1) If

$$\sup_{x\in[a,b]} |P(x)| \leq W$$

then for all x in (a, b) and for all i in $\mathbb{N}$, it holds;

THEOREM 6.2.– Furthermore,

$$\left|P^{(i)}(x)\right| = \frac{n^2\left(n^2-1\right)\ldots\left(n^2-(i-1)^2\right)}{(2i-1)!!}\left(\frac{2}{b-a}\right)^i W$$

if and only if

$$P(x) = W \cos\left(n \arccos \frac{2x - a - b}{b - a}\right)$$

i.e. when P is the Chebyshev polynomial of the first kind with degree n.

PROOF.– See Theorem A3.3 in Appendix 3. □

THEOREM 6.3 (Bernstein).– Let $P(x) := \sum_{j=0}^{n} a_j x^j$ be a polynomial with real coefficients and degree n. If $\sup_{x \in [a,b]} |P(x)| \leq W$, then for all x in (a, b) and for all i in $\mathbb{N}$, it holds

$$\left|P^{(i)}(x)\right| \leq \frac{n^2\left(n^2 - 1\right)\ldots\left(n^2 - (i-1)^2\right)}{(2i-1)!!}\left(\frac{2}{b-a}\right)^i W\left(1 + \mathcal{O}\left(\frac{1}{n^2}\right)\right).$$

PROOF.– See Theorem A3.4 in Appendix 3. □

6.6. Concentration inequalities

This section provides some crude bounds for deviation probabilities under mild hypotheses on the distribution of the underlying random variable. These describe the random term inherent to any measurement; two cases will be considered, according to the knowledge on this random term.

The first well-known inequality for events of the type

$$|X - E(X)| \geq \varepsilon,$$

for $\varepsilon > 0$, is the Chebyshev inequality, assuming that X has a finite variance. It holds

$$\Pr\{|X - E(X)| \geq \varepsilon\} \leq \frac{Var(X)}{\varepsilon^2}.$$

Strengthening the hypothesis on X leads to stronger inequalities for the deviation of X with respect to its expectation. We will make use of Hoeffding Inequality. We derive the generic result given by Markov.

PROPOSITION 6.1 (S. Markov).– Let X be a real valued random variable and t be a positive number. Then,

$$\Pr\{X \geq t\} \leq \frac{E(X^+)}{t} \leq \frac{E(|X|)}{t},$$

where $X^+ := \max\{X, \mathbf{0}\}$.

PROOF.– Clearly

$$\mathbf{I}_{[t,\infty)}(X) \leq \frac{X}{t}\mathbf{I}_{[t,\infty)}(X) \leq \frac{X^+}{t} \leq \frac{|X|}{t}.$$

Integration yields

$$\Pr\{X \geq t\} \leq \frac{1}{t}E\left(X^+\right) \leq \frac{1}{t}E\left(|X|\right).$$

REMARK 6.2.– When $X \in \mathbf{L}^p := \{X : \int |X|^p \, dP < \infty$ and $X = Y$ if $X = Y$ almost everywhere with respect to $P\}$ then $\Pr\{X \geq t\} \leq \frac{1}{t^p}E\left(|X|^p\right)$, for all $t > 0$. For any $t > 0$, $\{X \geq t\} \subset \{|X|^p \geq t^p\}$. When $X \in \mathbf{L}^2$, applying the Markov inequality to the RV $X - E(X)$ yields the Chebyshev inequality since for any $t > 0$, $\Pr\{|X - E(X)| \geq t\} \leq \frac{var(X)}{t^2}$.

PROPOSITION 6.2 (Hoeffding).– Let $\{X_i, i \in \mathbb{N}\}$ be a sequence of independent real valued random variables such that for any $i, X_i \in [a_i, b_i], (a_i, b_i) \in \mathbb{R}^2$. Let $S_n := \sum_{i=1}^{n} X_i$ and $\mu := \sum_{i=1}^{n} E(X_i)$. Then, for any $\varepsilon > 0$,

$$\Pr\{|S_n - \mu| \geq \varepsilon\} \leq 2\exp\left(-\frac{2\varepsilon^2}{\sum_{i=1}^{n}(b_i - a_i)^2}\right)$$

PROOF.– Let X be a RV with $E(X) = 0$ and $\Pr\{|X| > 1\} = 0$.We first prove that

$$E\left(e^{tX}\right) \leq \exp\left(\frac{t^2}{2}\right).$$

For any $|x| \leq 1$ and $t \in \mathbb{R}$, we write tx as a convex combination n the form

$$tx = \frac{1-x}{2}(-t) + \frac{1+x}{2}t.$$

Since the exponential function is convex, it follows that

$$e^{tx} \leq \frac{1-x}{2}\exp(-t) + \frac{1+x}{2}\exp(t),$$

which yields

$$e^{tX} \leq \frac{1-X}{2} \exp(-t) + \frac{1+X}{2} \exp(t).$$

Taking expectation (since X is a bounded r.v.)

$$E\left(e^{tX}\right) \leq \frac{1-E(X)}{2} \exp(-t) + \frac{1+E(X)}{2} \exp(t).$$

Now, $\Pr\{|X| > 1\} = 0$ and therefore, $0 \leq \frac{1-X}{2} \leq 1$ and $0 \leq \frac{1+X}{2} \leq 1$. It follows that

$$0 \leq E\left(\frac{1-X}{2}\right) \leq 1 \text{ and } 0 \leq E\left(\frac{1+X}{2}\right) \leq 1,$$

$$E\left(e^{tX}\right) \leq \frac{1-E(X)}{2} \exp(-t) + \frac{1+E(X)}{2} \exp(t)$$

$$\leq \exp(-t) + \exp(t) = cht := \sum_{j=0}^{\infty} \frac{t^{2j}}{(2j)!}.$$

Also by induction on n, it holds for any $n \in \mathbb{N}$, $n!2^n \leq (2n)!$. It follows that $cht \leq \exp\left(\frac{t^2}{2}\right)$. Hence,

$$E\left(e^{tX}\right) \leq \cosh t \leq \exp\left(\frac{t^2}{2}\right).$$

We now assume that we have a sequence or independent RVs $\{X_j, j \in \mathbb{N}\}$ such that for all j and $c_j > 0$

$$\Pr\{|X_j| \geq c_j\} = 0,\ E(X_j) = 0.$$

Since,

$$\Pr\left\{\left|\frac{X_j}{c_j}\right| \geq 1\right\} = 0, E(X_j) = 0,$$

we get, applying the above result, for any $r \in \mathbb{R}$

$$E\left(\exp\left(r\frac{X_j}{c_j}\right)\right) \leq \exp\left(\frac{r^2}{2}\right).$$

Choose $r = tc_j$. Then,

$$E\left(\exp\left(tX_j\right)\right) \leq \exp\left(\frac{t^2}{2}c_j^2\right).$$

The RVs $\exp(tX_j)$ are independent, which yields for all $t \in \mathbb{R}$,

$$E\left(\exp\left(t\sum_{j=1}^{n} X_j\right)\right) \leq \exp\left(\frac{t^2}{2}\sum_{j=1}^{n} c_j^2\right).$$

The mapping

$$x \mapsto \exp\left(tx\right),$$

is monotonous and non-decreasing. Hence for all $t > 0$, and all $\varepsilon > 0$, we have

$$\left\{\omega : \sum_{j=1}^{n} X_j\left(\omega\right) > \varepsilon\right\} \subset \left\{\omega : \exp\left(t\sum_{j=1}^{n} X_j\left(\omega\right)\right) > \exp\left(t\varepsilon\right)\right\}.$$

It follows that

$$\Pr\left\{\omega : \sum_{j=1}^{n} X_j\left(\omega\right) > \varepsilon\right\} \leq \Pr\left\{\omega : \exp\left(t\sum_{j=1}^{n} X_j\left(\omega\right)\right) > \exp\left(t\varepsilon\right)\right\}.$$

From the Markov inequality we have

$$\Pr\left\{\omega : \exp\left(t\sum_{j=1}^{n} X_j\left(\omega\right)\right) > \exp\left(t\varepsilon\right)\right\} \leq \frac{E\left(\exp\left(t\sum_{j=1}^{n} X_j\left(\omega\right)\right)\right)}{\exp\left(t\varepsilon\right)},$$

hence,

$$\Pr\left\{\omega : \sum_{j=1}^{n} X_j\left(\omega\right) > \varepsilon\right\} < \frac{E\left(\exp\left(t\sum_{j=1}^{n} X_j\left(\omega\right)\right)\right)}{\exp\left(t\varepsilon\right)}.$$

Since, $E\left(\exp\left(t\sum_{j=1}^{n} X_j\right)\right) \leq \exp\left(\frac{t^2}{2}\sum_{j=1}^{n} c_j^2\right)$, it follows that

$$\Pr\left\{\omega : \sum_{j=1}^{n} X_j\left(\omega\right) > \varepsilon\right\} < \frac{\exp\left(\frac{t^2}{2}\sum_{j=1}^{n} c_j^2\right)}{\exp\left(t\varepsilon\right)} = \exp\left(-t\varepsilon + \frac{t^2}{2}\sum_{j=1}^{n} c_j^2\right)$$

for all positive and ε. Optimize with respect to (t, ε), minimizing the RHS in the above display. We evaluate this minimal value. Denote $C := \sum_{j=1}^{n} c_j^2$, $\forall \varepsilon < 0$. The parabola

$$t \mapsto \frac{t^2}{2}C - t\varepsilon,$$

is minimal for $t = \frac{\varepsilon}{C}$ and assumes value $-\frac{\varepsilon^2}{2C}$.

Therefore, the minimal upper bound is $\exp\left(-\frac{\varepsilon^2}{2C}\right) = \exp\left(-\frac{\varepsilon^2}{2\sum_{j=1}^{n} c_j^2}\right)$. It follows that

$$\Pr\left\{\omega : \sum_{j=1}^{n} X_j(\omega) > \varepsilon\right\} < \exp\left(-\frac{\varepsilon^2}{2\sum_{j=1}^{n} c_j^2}\right).$$

We now obtain the Hoeffding inequality, noting that

$$\left\{\omega : \left|\sum_{j=1}^{n} X_j(\omega)\right| > \varepsilon\right\} = \left\{\omega : \sum_{j=1}^{n} X_j(\omega) > \varepsilon\right\} \cup \left\{\omega : -\sum_{j=1}^{n} X_j(\omega) > \varepsilon\right\},$$

$$P\left\{\omega : \left|\sum_{j=1}^{n} X_j(\omega)\right| > \varepsilon\right\} = P\left\{\omega : \sum_{j=1}^{n} X_j(\omega) > \varepsilon\right\}$$

$$+P\left\{\omega : -\sum_{j=1}^{n} X_j(\omega) > \varepsilon\right\}$$

$$= 2P\left\{\omega : \sum_{j=1}^{n} X_j(\omega) > \varepsilon\right\} < 2\exp\left(-\frac{\varepsilon^2}{2\sum_{j=1}^{n} c_j^2}\right).$$

6.7. Upper bounds of the extrapolation error due to randomness, and the resulting size of the design for real analytic regression functions

Consider an unknown real-valued analytic function f defined on some interval D:

$$f : \ D := (a, b) \to \mathbb{R}$$
$$v \mapsto f(v).$$

We assume that this function is observable on a compact subset S included in D, $S := [\underline{s}, \overline{s}] \subset D$, and that its derivatives are not observable at any point of D. The domain of extrapolation is denoted $U := D \setminus S$.

Let $X := \{s_k , k = 0, ..., l\}$ be the set of $l + 1$ Chebyshev nodes in S.

Let $Y_j (s_k)$ be the j-th observation on the knot s_k, $j = 1, \ldots, n_k$, $k = 0, \ldots, l$ and $n = \sum_{k=0}^{l} n_k$, $n_k > 0$ for all k. We suppose that the observation $Y_j (s_k)$ is generated according to the following regression model:

$$Y_j (s_k) = f (s_k) + \varepsilon_{j,k}, j = 1, ..., n_k, \quad k = 0, ..., l, \qquad [6.3]$$

where $\varepsilon_{j,k}$ is a completely specified continuous centered random variable.

We assume to observe $(l+1)$ IID samples, $\underline{Y} (k) := (Y_1 (n_k), ..., Y_{n_k} (n_k)), k = 0, ..., l$, and that for all $i \neq j$, $i = 0, \ldots, l$, $Y_i (n_k)$ and $Y_j (n_k)$ are independent.

The aim is to estimate $f^{(d)} (v)$ at any point $v < a$, hence on an extrapolation point.

Let

$$\mathcal{L}_l \left(f^{(d)}\right) (s) := \sum_{k=0}^{l} f (s_k) \, l_k^{(d)} (s)$$

be the Lagrange interpolation polynomial of the function $f^{(d)}$ with chosen degree l.

For all $d \in \mathbb{N}$, for any $v \in S$, the Lagrange interpolation scheme converges for the function $f^{(d)}$, that is, for $l \to \infty$; see Theorem 3.9,

$$\mathcal{L}_l \left(f^{(d)}\right) (s) \to f^{(d)} (s), \quad \text{for all} s \in S.$$

Interpolating the derivative $f^{(d+i)} (s^*)$ at a point $s^* \in S$ opportunely chosen, a Taylor expansion with order $(m-1)$ of $f^{(d)}(v)$ at point v from s^* gives

$$T_{f^{(d)},m,l} (v) := \sum_{i=0}^{m-1} \frac{(v - s^*)^i}{i!} \mathcal{L}_l \left(f^{(d+i)}\right) (s^*), \quad s^* \in S,$$

and we have

$$\lim_{m\to\infty} \lim_{l\to\infty} T_{f^{(d)},m,l} (v) = f^{(d)} (v), \quad \text{for all } v \in D.$$

The point s^* which designates where the Taylor expansion is derived from is free in the interpolation domain; since we intend to estimate $f^{(d)}(v)$ at some point outside the interpolation range, the smaller the distance between s^* and the frontier of the interpolation range, the smaller $M(m, l, \alpha)$ as defined in [6.4] hereunder, therefore leading to a better Taylor approximation. Additionally, s^* should minimize the error committed propagating the uncertainties due to the measurements on the knots, which is measured through the Lebesgue function $\lambda_{l+1}(X, s^*)$ where X designates the present Chebyshev nodes (see section 6.2); it is, however, known that the maximal value of the Lebesgue function is of order $\log l$ (see Proposition 3.2). This compromise is difficult to handle in practical cases; it is understandable to give some privilege to the first argument above.

When $f^{(d)} \in \mathcal{C}^\infty(D)$ (the set of all ∞-times continuously differentiable functions on D), then for any d and $l \geq 2d - 3$, the upper bound for the error of approximation is given in [BRO 12],

$$E_t := \sup_{v \in D} \left| f^{(d)}(v) - T_{f^{(d)},m,l}(v) \right| \leq M(m, l, \alpha),$$

where,

$$M(m, l, \alpha) = A(\alpha, l) + B(m) \tag{6.4}$$

with

$$A(\alpha, l) := K(\alpha, l) \sum_{i=0}^{m-1} \left(\sup_{s \in S} \left| f^{(d+i+\alpha)}(s) \right| \frac{1}{i!} \sup_{v \in U} \left| v - s^* \right|^i \right),$$

$$K(\alpha, l) := \left(\frac{\pi}{2(1+l)} (\overline{s} - \underline{s}) \right)^\alpha \left(9 + \frac{4}{\pi} \ln(1+l) \right),$$

and $B(m) := \sup_{v \in (a, \overline{s})} \left(\frac{|u - s^*|^m \left| f^{(d+\alpha)}(v) \right|}{m!} \right)$.

The optimal design is written $\{(n_k, s_k) \in (\mathbb{N} \setminus \{0\})^{l+1} \times \mathbb{R}^{l+1},\ n := \sum_{k=0}^{l} n_k,$ n fixed$\}$, where n is the total number of experiments and the $(l+1)$ knots are defined by

$$s_k := \frac{\overline{s} + \underline{s}}{2} - \frac{\overline{s} - \underline{s}}{2} \cos \frac{2k-1}{2l+2} \pi, \qquad k = 0, \ldots, l,$$

with $n_k := \left[\frac{n\sqrt{P_k}}{\sum_{k=0}^{l} \sqrt{P_k}} \right]$, where

$$P_k := \left| \sum_{\beta=0}^{m} \sum_{\alpha=0}^{m} \frac{(u-s)^{\alpha+\beta}}{\alpha!\beta!} l_k(s)\, l_k(s) \right|, \qquad k = 0, ..., l. \tag{6.5}$$

See [BRO 07] for details.

The function f cannot be observed exactly at the knots, due to randomness. Let $\widehat{f(s_k)}$ denote the least squares estimate of $f(s_k)$ at the knot s_k and define for all $i \geq 0$;

$$\mathcal{L}_l\left(\widehat{f^{(d+i)}}\right)(v) := \sum_{k=0}^{l} \widehat{f^{(d+i)}(s_k)} l_k(v). \qquad [6.6]$$

We estimate the d-th derivative of $f(v)$ at $v \in D$ as follows:

$$\widehat{T}_{f^{(d)},m,l}(v) := \sum_{i=0}^{m-1} \frac{(v-s^*)^i}{i!} \mathcal{L}_l\left(\widehat{f^{(d+i)}}\right)(s^*), \qquad s^* \in S.$$

The knots s_k are chosen to minimize the variance of $\widehat{T}_{f^{(d)},m,l}(v)$ and it holds (see [BRO 12])

$$\lim_{m\to\infty} \lim_{l\to\infty} \lim_{\min_{k=0,\ldots,l}(n_k)\to\infty} \widehat{T}_{f^{(d)},m,l}(u) = f^{(d)}(u), \text{ for all } u \in D. \qquad [6.7]$$

$\widehat{T}_{f^{(d)},m,l}(v)$ is an extrapolation estimator.

The result in [6.7] stems from the following arguments: the limit in $\min_{k=0,\ldots,l}(n_k) \to \infty$ is a consequence of the law of large numbers (least square estimator). The limit in $l \to \infty$, as the limit in $m \to \infty$, follows from the fact that f is an analytic function.

For a fixed degree l of the Lagrange scheme [6.6], the total error committed while substituting $f^{(d)}(v)$ by $\widehat{T}_{f^{(d)},m,l}(v)$ is written

$$E_{Tot}\left(f^{(d)}(v)\right) := f^{(d)}(v) - \widehat{T}_{f^{(d)},m,l}(v).$$

For the interpolation error concerning $f^{(i+d)}$, we have the following result presented in 3.10: since $f^{(i+d)} \in \mathcal{C}^{\infty}(S)$, for all d, and $l \geq 2d-3$, then

$$\sup_{s\in S}\left|f^{(d+i)}(s) - \mathcal{L}_l\left(f^{(d+i)}\right)(s)\right| \leq M_1 := K(d,l) \sup_{s\in S}\left|f^{(d+i)}(s)\right|.$$

This error depends on the choice of the knots and is controlled by the tuning of l.

The error due to the Taylor expansion of order $(m-1)$

$$f^{(d)}(v) - \sum_{i=0}^{m-1} \frac{(v-s^*)^i}{i!} f^{(d+i)}(s^*)$$

depends on s^*, is a truncation error and can be controlled by the tuning of m.

Let $\widehat{f(s_k)}$ be an estimate of $f(s_k)$ on the knot s_k and

$$e(k) := f(s_k) - \widehat{f(s_k)}, \qquad k = 0, ..., l$$

denote the error pertaining to $f(s_k)$ due to this estimation; this term $e(k)$ clearly depends on n_k, the frequency of observations at knot s_k.

Finally, when n is fixed, the error committed while extrapolating depends on the design $\{(n_k, s_k) \in (\mathbb{N} \setminus \{0\})^{l+1} \times \mathbb{R}^{l+1},\ k = 0, \ldots, l,\ n = \sum_{k=0}^{l} n_k\}$, on m and on l.

Without loss of generality, we will assume $\sigma = 1$, assumed to be known. In this case, we have

$$\widehat{f(s_k)} = \overline{Y}(s_k) := \frac{\sum_{j=1}^{n_k} Y_j(k)}{n_k}.$$

In the next section, we will provide upper bounds for the errors to restrict them.

Since f is supposed to be a real analytic function, we can consider the extrapolation as an analytic continuation of the function out of the set S obtained by a Taylor expansion from an opportunely chosen point s^* in S. So, the extrapolation error will depend on the order of the Taylor expansion and on the precision in the knowledge of the derivatives of the function at s^*. This precision is given by the interpolation error and by the estimation errors on the knots. The analyticity assumption also implies that the interpolation error will converge quickly to zero. For all integer r, the following result holds (see Theorem 3.9)

$$\lim_{l \to \infty} l^r \sup_{s \in S} \left| f^{(j)}(s) - \sum_{k=0}^{l} l_k^{(j)}(s) f(s_k) \right| = 0.$$

The extrapolation error depends on three kinds of errors: truncation error; interpolation error; error of estimation of the function on the knots. To control the extrapolation error, we split the upper bound for it in the sum of three terms, each term depending only on one of the three kinds of errors.

In the following we will distinguish two cases: in the first case, we suppose that the observed random variable Y is bounded, while Y is supposed to be a random variable with unbounded support in the second case. We suppose that the support is known.

6.7.1. *Case 1: the range of the observations is bounded*

If τ_1, τ_2 (assumed known), $\tau_1 \neq \tau_2$ are such that $\Pr(\tau_1 \leq Y \leq \tau_2) = 1$, it holds $|f(v)| \leq R$, where $R := \max\{|\tau_1|, |\tau_2|\}$. $E(Y) = f \in [-R, R]$. Let,

$$e(k) := \frac{\sum_{j=1}^{n_k} Y_j(k)}{n_k} - f(s_k).$$

The variables $Y_j(s_k)$, for all $j = 1, \ldots, n_k$, for all $k = 0, \ldots, l$, are IID, with the same bounded and for all k, $E(Y_j(s_k)) = f(s_k)$, hence we can apply Hoeffding's inequality; see [6.2].

$$\Pr\{|e(k)| \geq \rho\} \leq 2\exp\left(-\frac{2\rho^2 n_k}{(\tau_2 - \tau_1)^2}\right).$$

In Proposition 6.3, we identify the upper bound for the extrapolation error denoted by E_{ext}. This bound is the sum of the three terms, M_{Taylor}, controlling the error associated with the truncation of the Taylor expansion which defines $f^{(d)}$, M_{interp}, controlling the interpolation error and M_{est}, describing the estimation error on the knots.

It holds (see [BRO 14])

$$\left| f^{(d+i)}(s^*) - \sum_{k=0}^{l} l_k^{(d+i)}(s^*) f(s_k) \right| \leq M_{interp}$$

$$\left| f^{(d)}(v) - \sum_{i=0}^{m-1} \frac{f^{(d+i)}(s^*)}{i!}(v - s^*)^i \right| \leq M_{Taylor}$$

$$\left| \sum_{k=0}^{l} l_k^{(d+i)}(s^*) f^{(d)}(s_k) - \widehat{f^{(d+i)}(s^*)} \right| \leq M_{est}$$

where

$$M_{interp} := \frac{R}{(b-a)^{d+\alpha}} \sum_{i=0}^{m-1} \frac{|s^* - v|^i}{i!} K(l, \alpha) \frac{(d+i+\alpha)!}{(b-a)^i}, \qquad [6.8]$$

$$M_{Taylor} := R\frac{(d+m)!}{m!}\left|\frac{s^* - v}{b-a}\right|^m \frac{1}{(b-a)^d}, \tag{6.9}$$

and

$$M_{est} := \Lambda(l,m,d)\left(\max_{k=0,\ldots,l}|e(k)|\right).$$

In the latest display

$$\Lambda(l,m,d) := \sum_{i=0}^{m-1}\sum_{k=0}^{l}\frac{|s^* - v|^i}{i!}\left|l_k^{(d+i)}(s^*)\right|.$$

PROPOSITION 6.3.– For all $d, i \in \mathbb{N}\backslash\{0\}$, if $f^{(i+d)} \in \mathcal{C}^{d+i}(a,b)$, with $l \geq 2(d+i)-3$, then, for all $v \in U$,

$$|E_{ext}(v)| \leq M_{Taylor} + M_{interp} + M_{est}.$$

PROOF.– By using the Cauchy theorem on the derivatives of the analytic functions, we obtain

$$\begin{aligned}
\left|f^{(d)}(v) - \widehat{f^{(d)}(v)}\right| &= \Bigg|f^{(d)}(v) + \sum_{i=0}^{m-1}\frac{f^{(d+i)}(s^*)}{i!}(u-s^*)^i \\
&\quad - \sum_{i=0}^{m-1}\frac{f^{(d+i)}(s^*)}{i!}(v-s^*)^i - \widehat{f^{(d)}(v)}\Bigg| \\
&\leq \left|f^{(d)}(v) - \sum_{i=0}^{m-1}\frac{f^{(d+i)}(s^*)}{i!}(v-s^*)^i\right| + \left|\sum_{i=0}^{m-1}\frac{f^{(d+i)}(s^*)}{i!}(v-s^*)^i - \widehat{f^{(d)}(v)}\right| \\
&\leq \frac{\sup_{v\in U}\left|f^{(d+m)}(v)\right|}{m!}|s^*-v|^m + \Bigg|\sum_{i=0}^{m-1}\frac{f^{(d+i)}(s^*)}{i!}(v-s^*)^i \\
&\quad - \sum_{i=0}^{m-1}\frac{\widehat{f^{(d+i)}(s^*)}}{i!}(v-s^*)^i\Bigg| \\
&\leq \frac{R(m+d)!}{(b-a)^d m!}\left|\frac{s^*-v}{b-a}\right|^m + \left|\sum_{i=0}^{m-1}\frac{(v-s^*)^i}{i!}\left(f^{(d+i)}(s^*) - \widehat{f^{(d+i)}(s^*)}\right)\right|
\end{aligned}$$

$$\leq M_{Taylor} + \sum_{i=0}^{m-1} \frac{|s^* - v|^i}{i!} \left| f^{(d+i)}(s^*) - \widehat{f^{(d+i)}(s^*)} \right|$$

$$\leq M_{Taylor} + \sum_{i=0}^{m-1} \frac{|s^* - v|^i}{i!} \left| f^{(d+i)}(s^*) - \sum_{k=0}^{l} l_k^{(d+i)}(s^*) f(s_k) \right.$$
$$\left. + \sum_{k=0}^{l} l_k^{(d+i)}(s^*) f^{(d)}(s_k) - \widehat{f^{(d+i)}(s^*)} \right|$$

$$\leq M_{Taylor} + \sum_{i=0}^{m-1} \frac{(s^* - v)^i}{i!} \left| f^{(d+i)}(s^*) - \sum_{k=0}^{l} l_k^{(d+i)}(s^*) f(s_k) \right|$$
$$+ \sum_{i=0}^{m-1} \sum_{k=0}^{l} \frac{(s^* - v)^i}{i!} l_k^{(d+i)}(s^*) \left| f(s_k) - \overline{Y}(k) \right|$$

$$\leq M_{Taylor} + \sum_{i=0}^{m-1} \frac{|s^* - u|^i}{i!} K(l, \alpha) \left(\sup_{s \in S} \left| f^{(d+i+\alpha)}(s) \right| \right)$$
$$+ \sum_{i=0}^{m-1} \sum_{k=0}^{l} \frac{|s^* - u|^i}{i!} \left| l_k^{(d+i)}(s^*) \right| \left| f(s_k) - \overline{Y}(k) \right|$$

$$\leq M_{Taylor} + \frac{R}{(b-a)^{d+\alpha}} \sum_{i=0}^{m-1} \frac{|s^* - u|^i}{i!} K(l, \alpha) \frac{(d+i+\alpha)!}{(b-a)^i}$$
$$+ \sum_{i=0}^{m-1} \sum_{k=0}^{l} \frac{|s^* - v|^i}{i!} \left| l_k^{(d+i)}(s^*) \right| \left| f(s_k) - \overline{Y}(k) \right|$$

$$\leq M_{Taylor} + M_{interp} + \sum_{i=0}^{m-1} \sum_{k=0}^{l} \frac{|s^* - u|^i}{i!} \left| l_k^{(d+i)}(s^*) \right| \left| f(s_k) - \overline{Y}(k) \right| \leq$$

$$\leq M_{Taylor} + M_{interp} + \left(\max_{k=0,\ldots,l} |e(k)| \right) \sum_{i=0}^{m-1} \sum_{k=0}^{l} \frac{|s^* - v|^i}{i!} \left| l_k^{(d+i)}(s^*) \right|$$
$$= M_{Taylor} + M_{interp} + M_{est}. \qquad \square$$

The next proposition yields the smallest integer such that the error of estimation is not greater than a chosen threshold with a fixed probability.

PROPOSITION 6.4.– For all $\eta \in (0,1]$, fixing any $\rho \in \mathbb{R}^+$, there exists $n \in \mathbb{N}$, such that

$$Pr\left(\max_{k=0,\dots,l} |e(k)| \geq \frac{\rho}{\Lambda(l,m,d)}\right) \leq \eta. \quad [6.10]$$

PROOF.– If, for all k, it holds $|e(k)| \geq \frac{\rho}{\Lambda(l,m)}$, then $\max_{k=0,\dots,l} |e(k)| \geq \frac{\rho}{\Lambda(l,m)}$. We have,

$$\Pr\left(\max_{k=0,\dots,l} |e(k)| \geq \frac{\rho}{\Lambda(l,m,d)}\right) \leq \prod_{k=0}^{l} \Pr\left(|e(k)| \geq \frac{\rho}{\Lambda(l,m,d)}\right)$$

$$\leq \prod_{k=0}^{l} 2\exp\left(-\frac{2\rho^2}{(\Lambda(l,m,d))^2} n_k\right).$$

If we choose

$$n^* = \left[\frac{(l+1)\ln 2 - \ln \eta}{2}\left(\frac{\Lambda(l,m,d)(\tau_2-\tau_1)}{\rho}\right)^2\right], \quad [6.11]$$

the inequality

$$\eta \geq \prod_{k=0}^{l} 2\exp\left(-\frac{2\rho^2}{(\Lambda(l,m,d))^2} n_k\right)$$

holds. □

Proposition 6.5 gives the upper bound for the extrapolation error that depends on (l,m,n). We recall that the number of knots $l+1$ controls the interpolation error, m denotes the number of terms used in the Taylor expansion for $f^{(d)}$ and n is the total number of observations used to estimate $f(s_k), k = 0,..,l$. Hence, n controls the total estimation error.

PROPOSITION 6.5.– With the same hypotheses and notations, we have that, for all positive η there exists $(\rho_m, \rho_l, \rho_n) \in (\mathbb{R}^+)^3$, such that

$$|E_{ext}(v)| \leq \rho_m + \rho_l + \rho_n$$

holds with probability larger than η. Here ρ_m, ρ_l and ρ_n, on (m,l,n), among those three terms, only ρ_n is random.

PROOF.– The terms (M_{Taylor}, M_{interp}) are known functions of (m, l) following formulae [6.8] and [6.9], which therefore, can be found solving

$$(M_{Taylor}, M_{interp}) = (\rho_m, \rho_l) .$$

The statistician will impose given numerical values for ρ_m and ρ_l, i.e. for the maximal admissible bound for the error committed in the Taylor approximation, and respectively on the Lagrange interpolation error. Plug in those values of m and l in [6.11]. The value of ρ, in this formula, should be selected as the upper bound

$$\frac{\rho}{\Lambda(l, m, d)}$$

in the inequality [6.10] as required by the statistician. Henceforth, the optimal number of replications of the experiment is defined completely now, setting $\rho_n := \frac{\rho}{\Lambda(l,m,d)}$ and $n = n^*$, in accordance with equation [6.11]. Note that we have provided an explicit solution for the design. □

In the case of the estimation of $f(v)$ (i.e. when $d = 0$), we obtain for the couple (m, n) the explicit solution

$$m = \left[\frac{\ln \rho_m - \ln R}{\ln(s^* - v) - \ln(b - a)} \right],$$

$$n = \left[\frac{(l+1)\ln 2 - \ln \eta}{2} \left(\frac{\Lambda(l, m, 0)(\tau_2 - \tau_1)}{\rho} \right)^2 \right].$$

When $l \geq 2m - 3$, l is the solution of the equation

$$\rho_l = R \sum_{i=0}^{m-1} \frac{|s^* - v|^i}{i!} K(l, m) \frac{(i)!}{(b-a)^i}.$$

When estimating $f^{(d)}(v)$ for $d > 0$, the equations to be solved in the variables n, m and l are

$$m = \left[\frac{\ln \rho_m - \ln R}{\ln(s^* - v) - \ln(b - a)} \right],$$

$$n = \left[\frac{(l+1)\ln 2 - \ln \eta}{2} \left(\frac{\Lambda(l, m, d)(\tau_2 - \tau_1)}{\rho} \right)^2 \right]. \quad [6.12]$$

When $l \geq 2(d+m) - 3$, l is the solution of the equation

$$\rho_l = \frac{R}{(b-a)^d} \sum_{i=0}^{m-1} \frac{|s^* - v|^i}{i!} K(l,d) \frac{(i+d)!}{(b-a)^i}.$$

The term $\Lambda(l, m, d)$ imposes the evaluation of the various derivatives of the elementary Lagrange polynomials. Such calculation may be avoided, making use of a uniform upper bound of those, in accordance with Markov. This bound is to be inserted in [6.12], with an approached value for n^*.

THEOREM 6.4 (Markov).– Let $P(x) := \sum_{j=0}^{l} a_j s^j$ be a polynomial with real coefficients and degree l. If $\sup_{xs \in [a,b]} |P(x)| \leq W$, then for all x in (a,b) and for all i in $\mathbb{N}$, it holds

$$\left|P^{(j)}(x)\right| \leq \frac{l^2(l^2-1)\ldots\left(l^2-(j-1)^2\right)}{(2j-1)!!} \left(\frac{2}{(b-a)}\right)^j W.$$

When applied to the elementary Lagrange polynomial, it is readily checked that $W = \pi$. Clearly,

$$\left|l_k^{(d+i)}(s)\right| = \left|\frac{(-1)^k \sin\left(\frac{2k-1}{2l+2}\pi\right)}{l+1} \frac{\cos((l+1)\theta)}{\cos\theta - \cos\left(\frac{2k-1}{2l+2}\pi\right)}\right| \leq$$

$$\leq \frac{\left|\sin\left(\frac{2k-1}{2l+2}\pi\right)\right|}{l+1} \frac{|\cos((l+1)\theta)|}{\left|\cos\theta - \cos\left(\frac{2k-1}{2l+2}\pi\right)\right|} \leq$$

$$\leq \frac{\left|\sin\left(\frac{2k-1}{2l+2}\pi\right)\right|}{l+1} \frac{(l+1)\left|\theta - \frac{2k-1}{2l+2}\pi\right|}{\frac{1}{\pi}\sin\left(\frac{2k-1}{2l+2}\pi\right)\left|\theta - \frac{2k-1}{2l+2}\pi\right|} = \pi.$$

We used

$$|\cos((l+1)\theta)| = \left|\cos((l+1)\theta) - \cos\left((l+1)\frac{2k-1}{2l+2}\pi\right)\right|$$
$$\leq (l+1)\left|\theta - \frac{2k-1}{2l+2}\pi\right|$$

and $\cos\left((l+1)\frac{2k-1}{2l+2}\pi\right) = 0$. Moreover,

$$\left|\cos\theta - \cos\left(\frac{2k-1}{2l+2}\pi\right)\right| = 2\sin\left(\frac{\theta + \frac{2k-1}{2l+2}\pi}{2}\right)\left|\sin\left(\frac{\theta - \frac{2k-1}{2l+2}\pi}{2}\right)\right|.$$

The concavity of the sine function on $[0, \pi]$ implies

$$\sin\left(\frac{\theta + \frac{2k-1}{2l+2}\pi}{2}\right) \geq \frac{1}{2}\left(\sin\theta + \sin\left(\frac{2k-1}{2l+2}\pi\right)\right)$$

$$\left|\sin\left(\frac{\theta - \frac{2k-1}{2l+2}\pi}{2}\right)\right| \geq \frac{2}{\pi}\left|\theta - \frac{2k-1}{2l+2}\pi\right|, \theta \in [0, \pi].$$

REMARK 6.3.– The Cauchy theorem merely gives a rough upper bound. To obtain a sharper upper bound, we would assume some additional hypotheses on the derivatives of the function.

6.7.2. *Case 2: the range of the observations is unbounded*

If the support of the random variable Y is not bound and f is a polynomial of unknown degree t, $t \leq g-1$, with g known, it is still possible to give an upper bound for the estimation error. Since,

$$f^{(d)}(v) = \sum_{i=0}^{g-1} \frac{f^{(d+i)}(s^*)}{i!}(v - s^*)^i$$

$$= \sum_{i=0}^{g-1} \frac{\sum_{k=0}^{g-1} l_k^{(d+i)}(s^*) f(s_k)}{i!}(v - s^*)^i = \sum_{k=0}^{g-1} l_k^{(d)}(v) f(s_k),$$

$f^{(d)}(v)$ can be estimated as follows:

$$\widehat{f^{(d)}(v)} = \sum_{k=0}^{g-1} l_k^{(d)}(v) \overline{Y}(s_k).$$

It holds, setting $Var\varepsilon = 1$

$$Var\left(\widehat{f^{(d)}(v)}\right) = \sum_{k=0}^{g-1} \left(l_k^{(d)}(v)\right)^2 \frac{1}{n_k} \to 0,$$

Since $E\left(\widehat{f^{(d)}(v)}\right)$ converges to $f^{(d)}(v)$, convergence in probability holds. We use the Chebyshev's inequality to obtain the upper bound for the estimation error. For a given $\eta > 0$,

$$\Pr\left\{\left|\widehat{f^{(d)}} - f^{(d)}\right| \geq \eta\right\} \leq \frac{\sum_{k=0}^{g-1}\left(l_k^{(d)}(v)\right)^2 \frac{1}{n_k}}{\eta^2}.$$

If for all fixed ω, we aim to obtain

$$\Pr\left\{\left|\widehat{f^{(d)}} - f^{(d)}\right| \geq \eta\right\} \leq \omega$$

we can choose n^* as the integer part of the solution of the equation

$$\frac{\sum_{k=0}^{g-1}\left(l_k^{(d)}(u)\right)^2 \frac{V(\varepsilon)}{n_k}}{\eta^2} = \omega,$$

that is

$$n^* = \left[\frac{\sum_{k=0}^{g-1}\left(l_k^{(d)}(u)\right)^2 V(\varepsilon)}{\omega\eta^2}\right].$$

The integer n^* is such that the inequality $\Pr\left\{\left|\widehat{f^{(d)}} - f^{(d)}\right| \geq \eta\right\} \leq \omega$ is satisfied.

We remark that if we know the degree t of the polynomial, then it is sufficient to set $g-1=t$. When $d=0$, we have $n^* = \frac{\sum_{k=0}^{g-1}\left(l_k^{(d+i)}(u)\right)^2 V(\varepsilon)}{\omega\eta^2}$.

We underline that for $d=0$ and when t is known, $\widehat{f^{(d)}(v)} = \widehat{f(v)}$ coincides with Hoel's estimator.

Finally, if the support of the random variable is not bound and if f is analytic but not a polynomial, then we cannot apply the Cauchy theorem on the analytic functions. This is an intrinsic obstacle for this problem, without any solution if no further information on the model is assumed.

PART 3

Mathematical Material

Appendix 1

Normed Linear Spaces

A1.1. General notions

Unless specified, a space is not a linear space; for example, a metric space is a set equipped with a metric in the same way that a normed space is a normed set, not necessarily a linear normed space, etc.

This appendix reviews the basic definitions and properties of normed linear spaces. The main result assesses that the compactness of the unit sphere is equivalent to the finiteness of the dimension of the linear space, on which most of the approximation theory relies. Proposition A1.1 is fundamental to derive the topological properties of a normed linear space.

DEFINITION A1.1.– Let F be a non-void set. $(F, \|.\|_F)$ is a normed space if and only if the mapping $\|.\|_F \ : \ F^2 \ \rightarrow \ [0;\infty), f \ \mapsto \ \|f\|_F$, has the following properties:

1) $\|f\|_F = 0$ iff $f = 0$;

2) $\|\alpha f\|_F = |\alpha| \ \|f\|_F$, for any $\alpha \in \mathbb{R}$ and any $f \in F$;

3) $\|f + g\|_F \leq \|f\|_F + \|g\|_F$, for any $(f, g) \in F^2$.

By this definition, for any $f \in F$, $\|f\|_F \geq 0$. When $(F, +, \cdot)$ is a linear space, then F is a *normed linear (or vector) space*.

The norm generalizes some properties of the distance. Since the length of a segment is the distance between its end points, it comes in a natural way to introduce a metric on F, through the following definition of $d_{\|.\|_F}((f,g))$, the distance between f and g:

$$d_{\|.\|_F} : F^2 \to [0;\infty), (f,g) \mapsto d_{\|.\|_F}((f,g)) := \|f-g\|_F .$$

It is easy to verify that the mapping $d_{\|.\|_F}$ defines a metric. Henceforth, the space $\left(F, d_{\|.\|_F}\right)$ is a metric space.

This metric $d_{\|.\|_F}$ in turn induces a topology τ_F on F; the neighborhood $\mathcal{I}_f$ of f with radius $\epsilon \in [0;\infty)$and is defined by

$$\mathcal{I}_f(\epsilon) := \left\{g \in F : d_{\|.\|_F}(f,g) \leq \epsilon\right\}.$$

The space (F, τ_F) is therefore, a topological space.

Also, since on F we just have defined a distance d_F between points, we may also define the distance between any f in F and the null vector 0, and therefore, the norm induced by d_F is defined through

$$\|f\|_{d_F} := d_F(f,0).$$

We may now characterize the metric which is induced by a norm on a normed linear space. It holds:

PROPOSITION A1.1.– A metric d defined on a linear space F is induced by a norm defined on F if and only if

a) $d_F(f+z, g+z) = d_F(f,g)$, for any $f, g.z \in F$;

a) $d_F(\alpha f, \alpha g) = |\alpha|\, d_F(f,g)$, for any $\alpha \in \mathbb{R}$.

PROOF.– Assume that the metric d_F is induced by a norm $\|.\|_F$ defined on F. Then,

$d_F(f,g) := \|f-g\|_F$. Hence,

$$d_F(f+z, g+z) = \|(f+z)-(g+z)\|_F = \|f-g\|_F = d_F(f,g).$$

It also holds

$$d_F(\alpha f, \alpha g) := \|\alpha f - \alpha g\|_F = |\alpha|\, \|f-g\|_F = |\alpha|\, d_F(f,g).$$

Reciprocally assume that on F there exists a distance d_F with the following two properties:

a) $d_F(f+z, g+z) = d_F(f,g)$, for any $f, g, z \in F$;

b) $d_F(\alpha f, \alpha g) = |\alpha|\, d_F(f,g)$, for any $\alpha \in \mathbb{R}$.

Define $\|f\|_{d_F} := d_F(f, 0)$.

We prove that $f \to \|f\|_{d_F}$ is a norm. Properties 1) and 2) in Definition A1.1 clearly hold. For example, by b) we obtain

$$\|\alpha f\|_{d_F} := d_F(\alpha f, 0) = d_F(\alpha f, \alpha 0) = d_F(f, 0) = |\alpha|\, \|f\|_{d_F}, \text{ for any } \alpha \in \mathbb{R}.$$

The triangle inequality for $\|.\|_{d_F}$ is obtained by noting that from a) it holds $d_F(f + z, g+z) = d_F f, g)$, hence $d_F(f+g, 0) = d_F(f+g, 0+g = d_F(f+g, g)$.

Repeating this argument it holds

$$\begin{aligned}\|f+g\|_{d_F} &:= d_F(f+g, 0) = d_F(f+g, g) \le d_F(f, 0) + d_F(g, 0) \\ &= \|f\|_{d_F} + \|g\|_{d_F}.\end{aligned}$$ □

REMARK A1.1.– A metric which is induced by a norm is, therefore, characterized by the fact that it is translation invariant, namely $d_F(f+z, g+z) = d_F(f,g)$.

REMARK A1.2.– Any normed linear space is a metric space. The reciprocal statement does not hold. Consider the following example. Let (X, d_X) the metric space where

$$X := \{z_n \in \mathbb{C} : n \in \mathbb{N}\}$$

and

$$d_X\left((z_n)_n, (w_n)_n\right) := \sum_{n=1}^{\infty} \frac{1}{2^n} \frac{|z_n - w_n|}{1 + |z_n - w_n|}.$$

Recall that the triangle inequality of the metric d_X results from the action of the increasing function

$$t \mapsto g(t) := \frac{t}{1+t}$$

on both sides of the inequality $|x - z| \le |x - y| + |y - z|$.

Clearly $d_X\left(\alpha (z_n)_n, \alpha (w_n)_n\right) \neq |\alpha|\, d_X\left((z_n)_n, (w_n)_n\right)$. Therefore, the above proposition does not hold, and the metric cannot be induced by a norm.

REMARK A1.3.– Through the metric, various typical topological notions may be defined: convergent sequences, neighborhoods, border, limit sets, etc. A complete normed linear space is called a Banach space.

Among many neighborhoods, spheres play an important role.

DEFINITION A1.2.– Given a normed space $(F, \|.\|_F)$, an open sphere $S(f,r)$ with center $f \in F$ and with radius $r > 0$ is defined by

$$S(f,r) := \{g \in F : \|f-g\|_F < r\}.$$

The set

$$\overline{S}(f,r) := \{g \in F : \|f-g\|_F \leq r\}$$

is the closed sphere with center f and radius r.

PROPOSITION A1.2.– In a normed linear space $(F, \|.\|_F)$, closed spheres are convex sets.

PROOF.– Let

$$f, g \in \overline{S(h,r)} := \{q \in F : \|f-q\|_F \leq r, r > 0\}.$$

For any $\alpha \in [0;1]$. It holds that

$$\|h - \alpha f - (1-\alpha) g\|_F = \|\alpha h - \alpha f + (1-\alpha) h - (1-\alpha) g\|_F$$
$$\leq \alpha \|h - f\|_F + (1-\alpha) \|h-g\|_F \leq \alpha r + (1-\alpha) r = r.$$

Hence

$$(\alpha f + (1-\alpha) g) \in \overline{S(h,r)}.$$

An easy and importance property of the norm is the following. □

PROPOSITION A1.3.– Let $(F, \|.\|_F)$ be a normed space. Then,

$$| \, \|f\|_F - \|g\|_F \, | \leq \|f-g\|_F.$$

PROOF.– Indeed,

$$\|f\|_F = \|f - g + g\|_F \leq \|f-g\|_F + \|g\|_F.$$

Therefore,

$$\|f\|_F - \|g\|_F \leq \|f-g\|_F.$$

Substituting f by g and g by f

$$\|g\|_F - \|f\|_F \leq \|g - f\|_F = \|(-1)(g - f)\|_F = \|f - g\|_F .$$

The last inequality implies

$$-(\|f\|_F - \|g\|_F) \leq \|f - g\|_F$$

i.e.

$$\|f\|_F - \|g\|_F \geq -(\|f - g\|_F)$$

It follows that

$$-\|f - g\|_F \leq \|f\|_F - \|g\|_F \leq \|f - g\|_F .$$

i.e.

$$|\|f\|_F - \|g\|_F| \leq \|f - g\|_F . \qquad \square$$

PROPOSITION A1.4 (Continuity of the norm).– The mapping $\|.\|_F : (F, \|.\|_F) \to (\mathbb{R}, |.|), f \mapsto \|f\|_F$, is continuous

PROOF.– We prove that the norm is a continuous mapping at some generic point $g \in F$.

We, thus, have to prove that for any $f \in F$, where $\varepsilon > 0$, there exists some $\delta > 0$, such that $\|f - g\|_F < \delta$ implies $|\|f\|_F - \|g\|_F| < \varepsilon$.

Since $|\|f\|_F - \|g\|_F| \leq \|f - g\|_F < \delta$ it suffices to apply $\delta = \varepsilon$ and the claim is proved. $\square$

We now turn to a more general result.

PROPOSITION A1.5.– Let $(F_1, \|.\|_1)$ and $(F_2, \|.\|_2)$ be two linear normed spaces, and $T : F_1 \to F_2$ be a linear mapping. Then the four following statements are equivalent: 1) T is a Lipschitz mapping; 2) T is a continuous mapping; 3) T is continuous at $f := (\mathbf{0}, 4)$, where there exists $c \in [0, +\infty)$ such that for any $f \in F_1 : \|T(f)\|_2 \leq c\|f\|_1$.

PROOF.– Clearly, 1) $\Rightarrow$ 2) $\Rightarrow$3).

We prove that 3) $\Rightarrow$ 4). If T is continuous at $\mathbf{0}$, there exist some $\delta > 0$ such that for $f \in S(\mathbf{0},\delta)$ (which is a neighborhood of $\mathbf{0}$ with radius δ).

It holds that

$$T(S(\mathbf{0},\delta)) \subseteq S(T(\mathbf{0}),1) = S(\mathbf{0},1)$$

(since $T(\mathbf{0}) = \mathbf{0}$). We thus have,

$$f \in S_{\|.\|_1}(\mathbf{0},\delta) := \{f \in F_1 : \|f\|_1 < \delta\}$$

which implies that

$$T(f) \in S_{\|.\|_2}(\mathbf{0},1) := \{T(f) \in F_2 : \|T(f)\|_2 < 1\}.$$

This means that there exists $\delta > 0$, such that

$$\|f\|_1 < \delta \Rightarrow \|T(f)\|_2 < 1$$

Consider $f \in S(\mathbf{0},\delta) \setminus \{\mathbf{0}\}$; since $f \neq \mathbf{0}$, it holds $\|f\|_1 \neq 0$. Furthermore,

$$\frac{\delta f}{2\|f\|_1} \in S_{\|.\|_1}(\mathbf{0},\delta).$$

Clearly,

$$\left\|\frac{\delta f}{2\|f\|_1}\right\|_1 = \frac{\delta}{2}\left\|\frac{f}{\|f\|_1}\right\|_1 = \frac{\delta}{2} < \delta.$$

The image of $\frac{\delta f}{2\|f\|_1}$ through T then belongs to $S_{\|.\|_2}(\mathbf{0},1)$, i.e.

$$\left\|T\left(\frac{\delta f}{2\|f\|_1}\right)\right\|_2 < 1.$$

Denote

$$\frac{1}{c} := \frac{\delta}{2}.$$

Using the linearity of T it holds that

$$\frac{1}{c\|f\|_1}\|T(f)\|_2 < 1$$

i.e.

$$\|T(f)\|_2 < c\|f\|_1, \text{ for all } f \in S(\mathbf{0}, \delta) \setminus \{\mathbf{0}\}$$

which also holds true when $f = \mathbf{0}$.

We have proved that

$$\|T(f)\|_2 < c\|f\|_1, \text{ for all } f \in S(\mathbf{0}, \delta).$$

We will now prove that 4) $\Rightarrow$ 1).

Assume that there exists $c \in [0, +\infty)$ such that for any $f \in F_1 : \|T(f)\|_2 \leq c\|f\|_1$. We will prove that T is a Lipschitz mapping. For any $f, g \in F_1$,

$$\|T(f) - T(g)\|_2 = \|T(f-g)\|_2 \leq c\|f-g\|_1.$$

Hence, 1)$\Rightarrow$ 2)$\Rightarrow$3) and 4)$\Rightarrow$ 1). □

PROPOSITION A1.6.– Let $(F, \|.\|_F)$ be a normed linear space. Then, the mappings $T_1 : F^2 \to F, (f,g) \mapsto f+g$ and $T_2 : \mathbb{R} \times F \to F, (\alpha, f) \mapsto \alpha f$ are continuous. Furthermore, the mapping $F \to \mathbb{R}, f \mapsto \|f\|_F$ is a Lipschitz mapping.

PROOF.– T_1 and T_2 are linear mappings, hence continuous by the preceding theorem. Also, setting $c = 1$ in 4), it follows that the norm is a Lipschitz mapping. □

The fact that the linear mapping $T : (F_1, \|.\|_1) \to (F_2, \|.\|_2)$ is bound and may come bundled with some consequences. Consider the case when $T : \mathbb{R}^n \to \mathbb{R}^m$, $\mathbf{x} \mapsto T(\mathbf{x}) := A\mathbf{x}$, where A is some $m \times n$ matrix.

For $A \neq 0$, the image set by T is not a bounded set in $\mathbb{R}^n$. Noting,

$\|.\|_{\mathbb{R}^m}$ the Euclidian norm in $\mathbb{R}^m$

$$\|A(\alpha\mathbf{x})\|_{\mathbb{R}^m} = |\alpha| \|A\mathbf{x}\|_{\mathbb{R}^m} \to \infty, \text{ for } \alpha \to \infty.$$

However, the ratio

$$\frac{\|A\mathbf{x}\|_{\mathbb{R}^m}}{\|\mathbf{x}\|_{\mathbb{R}^n}},$$

is uniformly bounded. This gives rise to the following definitions:

DEFINITION A1.3.– Given two normed linear spaces $(F_1, \|.\|_1)$ and $(F_2, \|.\|_2)$ the operator (linear mapping) $T : (F_1, \|.\|_1) \rightarrow (F_2, \|.\|_2)$ is a bounded operator if for any $f \in F_1$, there exists some constant c such that $\|T(f)\|_2 \leq c \|f\|_1 \cdot$

The first definition yields the second one. Clearly, if T is linear and bound then when $f \neq \mathbf{0}$,

$$\|T(f)\|_2 \leq c\|f\|_1 \Rightarrow \frac{\|T(f)\|_2}{\|f\|_1} \leq c.$$

Taking the sup upon f in this display, it holds

$$\sup_{f \in F_1 \setminus \{\mathbf{0}\}} \frac{\|T(f)\|_2}{\|f\|_1} \leq c < +\infty.$$

Denote

$$\mathcal{L}(F_1, F_2) := \{T : F_1 \rightarrow F_2\}$$

where $(F_1, \|.\|_1)$ and $(F_2, \|.\|_2)$ are two normed linear spaces; then the mapping

$$\|.\|_{\mathcal{L}(F_1,F_2)} : \mathcal{L}(F_1, F_2) \rightarrow \mathbb{R}, T \mapsto \|T\|_{\mathcal{L}(F_1,F_2)} := |||T|||$$

is a norm.

Clearly,

$$\begin{aligned}
\|T\|_{\mathcal{L}(F_1,F_2)} = 0 &\Leftrightarrow \sup_{f \in F_1 \setminus \{\mathbf{0}\}} \frac{\|T(f)\|_2}{\|f\|_1} = 0 \\
&\Rightarrow 0 \leq \|T(f)\|_2 \leq |||T|||\ \|f\|_1 = 0, \text{ for all } f \in F_1, \\
&\Leftrightarrow \|T(f)\|_2 = 0, \text{ for all } f \in F_1.
\end{aligned}$$

Hence,

$$T = \mathbf{0}.$$

Furthermore,

$$\begin{aligned}
|||\alpha T||| = \sup_{f \in F_1 \setminus \{\mathbf{0}\}} \frac{\|\alpha T(f)\|_2}{\|f\|_1} &= \sup_{f \in F_1 \setminus \{\mathbf{0}\}} \frac{|\alpha|\ \|T(f)\|_2}{\|f\|_1} \\
= |\alpha| \sup_{f \in F_1 \setminus \{\mathbf{0}\}} \frac{\|T(f)\|_2}{\|f\|_1} &= |\alpha|\ |||T||| \cdot
\end{aligned}$$

Finally,

$$\begin{aligned} |||T_1 + T_2||| = & \sup_{f \in F_1 \setminus \{0\}} \frac{\|(T_1 + T_2)(f)\|_2}{\|f\|_1} \\ = & \sup_{f \in F_1 \setminus \{0\}} \frac{\|T_1(f) + T_2(f)\|_2}{\|f\|_1} \\ \leq & \sup_{f \in F_1 \setminus \{0\}} \frac{\|T_1(f)\|_2}{\|f\|_1} + \sup_{f \in F_1 \setminus \{0\}} \frac{\|T_2(f)\|_2}{\|f\|_1} \end{aligned}$$

$$= |||T_1||| + |||T_2||| .$$

This yields the following definition.

DEFINITION A1.4.– The real number

$$|||T||| := \sup_{f \in F_1 \setminus \{0\}} \frac{\|T(f)\|_2}{\|f\|_1}$$

is called the norm of the operator T.

PROPOSITION A1.7.– Let $(F_1, \|.\|_1)$ and $(F_2, \|.\|_2)$ be two normed linear spaces. Then, $\left(\mathcal{L}(F_1, F_2), \|.\|_{\mathcal{L}(F_1,F_2)}\right)$ is a normed linear space.

PROOF.– It follows from the above arguments. □

Recall that in $\mathbb{R}^n$ all norms are equivalent, i.e. they induce the same topology. This also holds true in any finite dimensional linear space.

When V is a linear space on $\mathbb{R}$ with dimension n, then V is isomorphic to $\mathbb{R}^n$. Therefore, $\mathbb{R}^n$ and V are indistinguishable as seen from the vector standpoint. Let a norm be defined on V. We consider the induced topology on V and ask whether the topological properties on $\mathbb{R}^n$ and V can be distinguished.

For example, since on $\mathbb{R}^n$, a subset $M \subset \mathbb{R}^n$ is compact if it is closed and bound, we may expect that the same characterization holds for compact subsets in V.

This holds true and we will prove it after some preliminary simple results.

We begin by proving the intuitive fact that combining in a linear way independent vectors with "large" weights, we obtain a vector which cannot have a "small" norm.

PROPOSITION A1.8.– Let $(F, \|.\|_F)$ be a normed linear space. Let $\{\varphi_1, ..., \varphi_n\} \subset F$ be a family of n independent vectors. Then, for any uple, $(\alpha_1, ..., \alpha_n)' \in \mathbb{R}^n$, there exists a real number c, such that

$$\left\| \sum_{i=1}^{n} \alpha_i \varphi_i \right\|_F \geq c \sum_{i=1}^{n} |\alpha_i| .$$

PROOF.– The claim surely holds if $y := \sum_{i=1}^{n} \alpha_i \varphi_i$ is the null vector. In this case $\|\sum_{i=1}^{n} \alpha_i \varphi_i\|_F = 0$, and also $\sum_{i=1}^{n} |\alpha_i| = 0$ (i.e. $\alpha_i = 0$, for all $i = 1, ..., n$).

The fact that the weights α_i assume value 0 stems from the fact that y is defined as a linear combination of the independent vectors φ_i. Therefore for any c,

$$\|y\|_F = \left\| \sum_{i=1}^{n} \alpha_i \varphi_i \right\|_F = 0 \geq c \sum_{i=1}^{n} |\alpha_i| = 0$$

holds true.

Consider the case $\sum_{i=1}^{n} |\alpha_i| > 0$.

Then,

$$\left\| \sum_{i=1}^{n} \alpha_i \varphi_i \right\|_F \geq c \sum_{i=1}^{n} |\alpha_i|$$

if

$$\left\| \sum_{i=1}^{n} \left(\frac{\alpha_i}{\sum_{i=1}^{n} |\alpha_i|} \right) \varphi_i \right\|_F \geq c.$$

For any i, $\beta_i := \frac{\alpha_i}{\sum_{i=1}^{n} |\alpha_i|}$; we prove that for any $\sum_{i=1}^{n} \beta_i \varphi_i$ with $\sum_{i=1}^{n} |\beta_i| = 1$, there exists some c such that

$$\left\| \sum_{i=1}^{n} \beta_i \, \varphi_i \right\|_F \geq c.$$

Assume, by contradiction, that this does not hold.

Then for any $c > 0$, for some $n-$uple $\beta := (\beta_1, ..., \beta_n)' \in \mathbb{R}^n$ with $\sum_{i=1}^{n} \beta_i = 1$, we have

$$\left\| \sum_{i=1}^{n} \beta_i \, \varphi_i \right\|_F < c.$$

Consider for example $c = \frac{1}{m}$, with m not zero.

Define a vector $y^{(m)}$ such that

$$\left\| y^{(m)} \right\|_F < \frac{1}{m},$$

where $y^{(m)} := \sum_{i=1}^{n} \beta_i^{(m)} \varphi_i$, and $\sum_{i=1}^{n} \left| \beta_i^{(m)} \right| = 1$.

Since,

$$\left\| y^{(m)} \right\|_F < \frac{1}{m}$$

holds for all m, taking the limit as $m \to \infty$,

$$\lim_{m \to \infty} \left\| y^{(m)} \right\|_F \to 0 \text{ which yields } \lim_{m \to \infty} y^{(m)} \to 0.$$

Since for any m, we have

$$\sum_{i=1}^{n} \left| \beta_i^{(m)} \right| = 1,$$

it follows that for all $i = 1, ..., n$, and any $m \in \mathbb{N} \setminus \{0\}$,

$$\left| \beta_i^{(m)} \right| \leq 1.$$

This means that the following n sequences

$$\left\{ \beta_1^{(m)} : m \in \mathbb{N} \setminus \{0\} \right\} = \left\{ \beta_1^{(1)}, ..., \beta_1^{(m)}, ... \right\}$$

....

$$\left\{ \beta_i^{(m)} : m \in \mathbb{N} \setminus \{0\} \right\} = \left\{ \beta_i^{(1)}, ..., \beta_i^{(m)}, ... \right\}$$

....

$$\left\{ \beta_n^{(m)} : m \in \mathbb{N} \setminus \{0\} \right\} = \left\{ \beta_n^{(1)}, ..., \beta_n^{(m)}, ... \right\}$$

are all bounded sequences.

Through Bolzano–Weierstrass we may extract from each of them a convergent subsequence.

Let, for any $i = 1, ..., n,$

$$\left\{\beta_i^{(r_m)} : r_m \in \mathbb{N} \setminus \{0\}\right\} = \left\{\beta_i^{(r_1)}, ..., \beta_i^{(r_m)}, ...\right\},$$

and

$$\lim_{r_m \to \infty} \beta_i^{(r_m)} = \lambda_i.$$

We may further build subsequences $\left\{y^{(m)} : m \in \mathbb{N} \setminus \{0\}\right\}$, defining

$$y_{i,m} := \sum_{j=1}^{n} \beta_j^{(r_m)} \varphi_j,$$

for $i = 1, ..., n.$

Let y be a linear combination of the $\varphi_i' s$ with coefficients the limiting values of the sequences $\left\{\beta_j^{(r_m)}\right\}$ .Then,

$$y := \sum_{j=1}^{n} \lambda_i \varphi_j.$$

Consider $\lim_{m \to \infty} \left\| y_{i,m} - \sum_{j=1}^{n} \lambda_j \varphi_j \right\|_F.$

We have

$$\left\| y_{i,m} - \sum_{j=1}^{n} \lambda_j \varphi_j \right\|_F = \left\| \sum_{j=1}^{n} \beta_j^{(r_m)} \varphi_j - \sum_{j=1}^{n} \lambda_j \varphi_j \right\|_F = \left\| \sum_{j=1}^{n} \left(\beta_j^{(r_m)} - \lambda_j\right) \varphi_j \right\|_F$$

$$= \sum_{j=1}^{n} \left|\beta_j^{(r_m)} - \lambda_j\right| \|\varphi_j\|_F.$$

It follows that

$$\lim_{m \to \infty} \left\| y_{i,m} - \sum_{j=1}^{n} \lambda_j \varphi_j \right\|_F = \lim_{m \to \infty} \left(\sum_{j=1}^{n} \left|\beta_j^{(r_m)} - \lambda_j\right| \|\varphi_j\|_F \right)$$

$$= \|\varphi_j\|_F \left(\sum_{j=1}^{n} \lim_{m \to \infty} \left|\beta_j^{(r_m)} - \lambda_j\right| \right)$$

$$= \|\varphi_j\|_F \left(\sum_{j=1}^{n} \left| \lim_{m\to\infty} \beta_j^{(r_m)} - \lambda_j \right| \right) = \|\varphi_j\|_F \left(\sum_{j=1}^{n} |\lambda_j - \lambda_j| \right) = 0.$$

Hence

$$\lim_{m\to\infty} \left\| y_{i,m} - \sum_{j=1}^{n} \lambda_j \varphi_j \right\|_F = 0.$$

We obtained a subsequence $y_{i,m}$ from y_m, which converges to $y = \sum_{j=1}^{n} \lambda_j \varphi_j$, where $\sum_{j=1}^{n} |\lambda_j| = 1$ and the $\varphi_j's$ are linearly independent.

Since $y_m \to 0$ as $m \to \infty$, any subsequence $\{y_{i,m}\}$ from $\{y_m\}$ converges to the null vector. Therefore, $y = 0$.

Now, by independence, of the φ_j, $j = 1, ..., n$, we conclude that $\lambda_j = 0$ for all j.

Also, since $\sum_{i=1}^{n} \left|\beta_i^{(m)}\right| = 1$ for all m, there exists at least one index i with $\beta_i^{(m)} \neq 0$.

We proved the claim, since then $\beta_1^{(m)} = ... = \beta_i^{(m)} = ... = \beta_n^{(m)} = 0$ and $\beta_i^{(m)} \neq 0$. □

The above result can derive some remarkable consequences.

THEOREM A1.1.– Any finite dimensional linear subspace V of a normed linear space F is complete.

PROOF.– Consider a Cauchy sequence in V. We prove that it converges in V.

Let $\dim V = n$, and $\{\varphi_1, ..., \varphi_n\}$ be a basis in V, and $\{y_m \in V : m \in \mathbb{N}\}$ be a Cauchy sequence in V. We can also write

$$y_m = \sum_{i=1}^{n} \alpha_{i,m} \varphi_i.$$

By definition of a Cauchy sequence for any $\varepsilon > 0$, there exists $h(\varepsilon) \in \mathbb{N}$, such that for any $m > h$, for any $j > h$, it holds $\|y_m - y_j\|_F \leq \varepsilon$.

Applying the preceding proposition to the vector $y_m - y_j := \sum_{i=1}^{n} (\alpha_{i,m} - \alpha_{i,j}) \varphi_i$, we obtain that there exists some $c > 0$, such that

$$\|y_m - y_j\|_F \geq c \sum_{i=1}^{n} |\alpha_{i,m} - \alpha_{i,j}|.$$

Together with the Cauchy property, for any $m, j > h(\varepsilon) \in \mathbb{N}$,

$$c\sum_{i=1}^{n} |\alpha_{i,m} - \alpha_{i,j}| \leq \|y_m - y_j\|_F \leq \varepsilon.$$

Henceforth, since $c > 0$,

$$\sum_{i=1}^{n} |\alpha_{i,m} - \alpha_{i,j}| \leq \frac{\varepsilon}{c}.$$

Now $|\alpha_{i,m} - \alpha_{i,j}| \leq \sum_{i=1}^{n} |\alpha_{i,m} - \alpha_{i,j}|$, which entails that for any $m, j > h(\varepsilon) \in \mathbb{N}$,

$$|\alpha_{i,m} - \alpha_{i,j}| \leq \frac{\varepsilon}{c}.$$

Hence, the n sequences $\{\alpha_{i,m} : m \in \mathbb{N}\}$, $i = 1, ..., n$, are Cauchy sequences in $\mathbb{R}$. Since $\mathbb{R}$ is a complete space, it follows that as $m \to \infty$, and for all $i = 1, ..., n$, $\alpha_{i,m} \to \alpha_i \in \mathbb{R}$.

Through these n limits, $\alpha_1, ..., \alpha_n$, we build a vector

$$y := \sum_{i=1}^{n} \alpha_i \varphi_i.$$

Clearly, $y \in V$. Furthermore,

$$\|y_m - y_j\|_F = \left\| \sum_{i=1}^{n} \alpha_{i,m}\varphi_i - \sum_{i=1}^{n} \alpha_i \varphi_i \right\|_F = \left\| \sum_{i=1}^{n} (\alpha_{i,m} - \alpha_i)\varphi_i \right\|_F$$

$$= \sum_{i=1}^{n} |\alpha_{i,m} - \alpha_i| \, \|\varphi_i\|_F \, .$$

Going to the limit in m,

$$\lim_{m\to\infty} \|y_m - y\|_F = \sum_{i=1}^{n} \left(\lim_{m\to\infty} |\alpha_{i,m} - \alpha_i| \right) \|\varphi_i\|_F = 0.$$

Therefore, the Cauchy sequence $\{y_m : m \in \mathbb{N}\} \subset V$ converges to $y \in V$. □

To characterize closed sets in a linear normed space, we derive a result on the closure of metric spaces.

PROPOSITION A1.9.– Let (E, d) be a metric space. Let $C \neq \phi$ be a subset in E. Let $\overline{C}$ be the closure of C (a subset in E which contains all limit points of C). Then:

1) $y \in \overline{C}$ if y is a limit of a sequence in C;

2) C is closed if all limit points of sequences in C belong to C.

PROOF.– To prove 1), consider $y \in \overline{C}$ and consider the two possible cases: a) $y \in C$, b) $y \notin C$.

In case a), clearly $y = \lim_{n\to\infty} y$. Now, any point in this constant sequence belongs to C.

In case b), let $y = \lim_{n\to\infty} x_n$ with $x_n \in C$ for all n and $y \notin C$.

Now, y is a limit point in C ; therefore, any of its neighborhoods contains points in C.

Consider the following neighborhoods of y : $S\left(y, \frac{1}{n}\right)$, $n \in \mathbb{N} \setminus \{0\}$, the sphere with center y and radius $1/n$. Hence, $S\left(y, \frac{1}{n}\right) \cap C \neq \phi$.

For $n = 1$, choose a point $x_1 \in S(y, 1) \cap C$, for $n = 2$, consider $x_2 \in S\left(y, \frac{1}{2}\right) \cap C$, for n the point $x_n \in S\left(y, \frac{1}{n}\right) \cap C$ etc.

As $n \to \infty$, $S\left(y, \frac{1}{n}\right) \to \{y\}$ and the sequence $x_1, ..., x_n$ converges to y.

The reciprocal statement holds as follows:

Let $\{x_n \in C : n \in \mathbb{N}\}$ be a sequence in C which converges to y. Two cases hold: a) $y \in C$ and then the claim is obvious; b) $y \notin C$. Then, y is a cluster point for $\{x_n \in C : n \in \mathbb{N}\} \subset C$ and therefore, for C. Hence, $y \in \overline{C}$. □

Claim b) holds since C is closed if and only if $C = \overline{C}$.

We quote the following consequence.

THEOREM A1.2.– Any finite dimensional linear subspace V of a normed linear space F is closed in F.

PROOF.– It is enough to observe that V is complete, by the previous proposition. □

Some norms induce the same topology. They are denominated as equivalent norms. The following definition makes this notion more precise:

DEFINITION A1.5.– Let F be a normed space equipped with the two norms: $\|.\|_1$ and $\|.\|_2$. These norms are equivalent if there exist $\alpha > 0$ and $\beta > 0$, such that for any $f \in F$, it holds $\alpha \|f\|_2 \leq \|f\|_1 \leq \beta \|f\|_2$.

REMARK A1.4.– In a normed space $(F, \|.\|_1)$ substitute $\|.\|_1$ by $\|.\|_2$ in a norm equivalent to $\|.\|_1$. Then, all notions pertaining to the norm $\|.\|_1$ hold with respect to $\|.\|_2$; the two norms define the same topology.

The following result is important.

PROPOSITION A1.10.– In a normed linear space V with finite dimension, all norms are equivalent.

PROOF.– Let $\|.\|_1$ and $\|.\|_2$ be two norms defined on V.

Let $\{\varphi_1, ..., \varphi_n\}$ be a basis of V.

For any $v \in V$, denote $v = \sum_{i=1}^n \alpha_i \varphi_i$.

$$\sum_{i=1}^n |\alpha_i| \leq \frac{\|\sum_{i=1}^n \alpha_i \varphi_i\|_1}{c}. \quad \text{[A1.1]}$$

Furthermore,

$$\left\| \sum_{i=1}^n \alpha_i \varphi_i \right\|_2 \leq \sum_{i=1}^n |\alpha_i| \, \|\varphi_i\|_2 \leq (\max \|\varphi_i\|_2) \sum_{i=1}^n |\alpha_i| \, .$$

Hence, by [A1.1]

$$\left\| \sum_{i=1}^n \alpha_i \varphi_i \right\|_2 \leq (\max \|\varphi_i\|_2) \sum_{i=1}^n |\alpha_i| \leq (\max \|\varphi_i\|_2) \frac{\|\sum_{i=1}^n \alpha_i \varphi_i\|_1}{c}.$$

Therefore,

$$\frac{c}{\max \|\varphi_i\|_2} \|y\|_2 \leq \|y\|_1 \, .$$

Set

$$\alpha = \frac{c}{\max \|\varphi_i\|_2} > 0,$$

to obtain

$$\alpha \|y\|_2 \leq \|y\|_1 \, .$$

The second inequality

$$\|y\|_1 \leq \beta \|y\|_2 ,$$

follows through similar arguments. □

REMARK A1.5.– The preceding proposition does not hold in infinite dimensional normed linear spaces. Let $F = \mathcal{C}^{(0)}([0;1])$, $\|f\|_1 := \int_0^1 |f(x)|\, dx$ and $\|f\|_2 := \sup_{[0:1]} |f(x)|$. We look for two non-negative constants α and β such that $\alpha \|f\|_1 \leq \|f\|_2 \leq \beta \|f\|_1$. Clearly

$$\int_0^1 |f(x)|\, dx = \|f\|_1 \leq \int_0^1 \left(\sup_{[0;1]} |f(x)| \right) dx$$
$$= \left(\sup_{[0;1]} |f(x)| \right) \int_0^1 dx = \sup_{[0;1]} |f(x)| = \|f\|_2 .$$

Therefore, the first inequality holds with $\alpha = 1$. We search for $\beta > 0$ such that

$$\|f\|_2 \leq \beta \|f\|_1 .$$

Consider the function

$$f_n(x) := \begin{cases} 1 - nx \text{ if } x \in \left[0 : \frac{1}{n}\right] \\ 0 \qquad\quad \text{if } x \in \left(\frac{1}{n}; 1\right] \end{cases} .$$

For any n,

$$\|f_n\|_2 = \sup_{[0;1]} |f_n(x)| = 1$$

and

$$\|f\|_1 = \int_0^1 |f_n(x)|\, dx = \int_0^{\frac{1}{n}} |1 - nx|\, dx = \frac{1}{2n}.$$

This implies $1 \leq \beta \frac{1}{2n}$, and therefore, for all n,

$$\beta \geq 2n$$

which proves that no such β exists.

REMARK A1.6.– The equivalence of norms implies that the asymptotic behavior of a function which belongs to a finite dimensional linear space does not depend on the norm.

We now characterize compactness in finite dimensional normed linear spaces.

Recall that metric space (E, d) is compact if from any sequence of elements in E, we may extract a convergent subsequence. A subset M in E is compact if it is as compact as a metric space.

A necessary condition for compactness is the following:

PROPOSITION A1.11.– A compact subset M in a metric space (E, d) is closed and bounded.

PROOF.– By definition since M is compact any sequence in M admits a convergent subsequence which converges to a point in M. Let $\overline{M}$ be the closure of M. For any x in $\overline{M}$ there exists a sequence $\{x_n, n \in \mathbb{N}\} \subset M$ such that $x_n \to x \in \overline{M}$. Since M is compact, x belongs to M.

We proved that M is bound, or else its diameter is infinite,

$$\sup_{(x,y)\in M^2} d(x, y) = \infty.$$

This entails that for any fixed a in M, $\sup_{y\in M} d(y, a) = \sup_{x\in M} d(x, a) = +\infty$; otherwise

$$\begin{aligned} d(x, y) &\leq d(x, a) + d(a, y) \\ &\Rightarrow \sup_{(x,y)\in M^2} d(x, y) \leq \sup_{x\in M} d(x, a) + \sup_{y\in M} d(y, a) < \infty \end{aligned}$$

holds; a contradiction.

Hence, supposing M unbound $\sup_{y\in M} d(y, a) = \infty$ for all a. It follows that there exists a sequence $\{x_n \in M : n \in \mathbb{N}\}$ such that $d(x_n, a) > n$ which yields $\lim_{n\to\infty} d(x_n, a) = \infty$. From this sequence, no convergent subsequence can be extracted. Hence, M is not compact, which proves that M should be bound. □

REMARK A1.7.– The reciprocal statement of this proposition usually does not hold. Let $E := l^2$, the class of all real valued sequences with finite square norm, and let $d(\{x_n\}, \{y_n\}) := \|\{x_n - y_n\}\|_{l^2} := \sqrt{\sum_{n=1}^{\infty} (x_n - y_n)^2}$. Consider the sequence x_n then defined through

$$x_n(i) := \begin{cases} 1^i \text{ if } n = i \\ 0^n \text{ if } n \neq i \end{cases}$$

for some fixed i. This sequence is bounded since $\|\{x_n(i)\}\|_{l^2} = 1$. For $i \neq j$, $\|\{x_n(i) - x_n(j)\}\|_{l^2} = \sqrt{2}$. This sequence is a closed set in $\mathbb{R}$ as a countable family of isolated points and is therefore, a closed set. However, since for $i \neq j$, $\|\{x_n(i) - x_n(j)\}\|_{l^2} = \sqrt{2}$, this sequence has no cluster point; E therefore is not compact.

In the above example E is an infinite dimensional linear space. We may ask if this makes the closeness and boundedness a non-sufficient condition for compactness. This is the case, as seen now.

THEOREM A1.3.– In a linear normed space on $\mathbb{R}$ with finite dimension $(V, \|.\|_V)$, M is a bounded and closed set if and only if it is compact.

PROOF.– If M is a compact set then it is bound and closed (see the above proposition A1.11).

Reciprocally assume that M is closed and bound, and let $\{\varphi_1, ..., \varphi_n\}$ be a basis in V.

Let $\{f_n \in M : n \in \mathbb{N}\}$ be a sequence in M. Then, $f_n := \sum_{i=1}^{n} \alpha_{i,m}\varphi_i$. Since $\{f_n \in V : n \in \mathbb{N}\} \subset M$, this sequence is bound. Therefore, there exists k such that for any n, $\|f_n\|_V \leq k$. Furthermore, there exists some $c > 0$, such that

$$\|f_n\|_V = \left\|\sum_{i=1}^{n} \alpha_{i,m}\varphi_i\right\|_V \geq c\sum_{i=1}^{n} |\alpha_{i,m}| .$$

Hence,

$$c\sum_{i=1}^{n} |\alpha_{i,m}| \leq \left\|\sum_{i=1}^{n} \alpha_{i,m}\varphi_i\right\|_V = \|f_n\|_V < k.$$

Henceforth, for any fixed i,

$$\sum_{i=1}^{n} |\alpha_{i,m}| < \frac{k}{c},$$

which implies that $|\alpha_{i,m}| < \frac{k}{c}$ for any $i = 1, ..., n$.

For any $i = 1, ..., n$, the sequence $\{\alpha_{i,m} \in \mathbb{R} : m \in \mathbb{N}\}$ is bounded. According to the Bolzano–Weierstrass theorem, it results that it has a cluster point.

We, thus, may extract a convergent subsequence $\{\alpha_{i,r_m} \in \mathbb{R} :, r_m \in \mathbb{N}\}$. Let λ_i be its limit as $r_m \to \infty$.

We may proceed in the same way for any coefficient $\alpha_{i,m}$ of the sequence f_n. Henceforth, we may extract from $\{f_n \in M : n \in \mathbb{N}\}$ a convergent subsequence with limit

$$f := \sum_{i=1}^{n} \left(\lim_{r_m \to \infty} \alpha_{i,r_m} \right) \varphi_i = \sum_{i=1}^{n} \lambda_i \varphi_i.$$

Since M is closed, f belongs to M. This proves that M is a compact set. □

THEOREM A1.4 (Riesz).– Let $(F, \|.\|_F)$ be a normed linear space and V a linear sub-space of F. If there exists some k, $0 < k < 1$, such that for any f in F satisfying $\|f\|_F = 1$ it holds $d_{\|.\|_F}(f, V) < k$, then V is a dense subset of F.

PROOF.– By contrast, suppose that there exists $0 < k < 1$, such that $d_{\|.\|_F}(f, V) < k$, for any f in F which satisfies $\|f\|_F = 1$, and assume that V is not dense in F.

Since, V is not dense its closure $\overline{V}$ does not coincide with F. Hence, $\overline{V} \subsetneq F$.

Consider a vector $f \in F \diagdown V$. Then, $d_{\|.\|_F}\left(f, \overline{V}\right) = \varepsilon > 0$. By definition

$$d_{\|.\|_F}\left(f, \overline{V}\right) := \inf_{g \in \overline{V}} d_{\|.\|_F}(f, g).$$

Hence, since $d_{\|.\|_F}\left(f, \overline{V}\right) = \varepsilon$, there exists some $g^* \in \overline{V}$, such that $\varepsilon = \inf_{g \in \overline{V}} d_{\|.\|_F}(f, g) = d_{\|.\|_F}(f, g^*) = \|f - g^*\|_F$.

Therefore,

$$\varepsilon \le \|f - g^*\|_F,$$

and since $0 < k < 1$,

$$\|f - g^*\|_F \le \frac{\varepsilon}{k}.$$

We thus have

$$\varepsilon \le \|f - g^*\|_F \le \frac{\varepsilon}{k}.$$

Consider the vector $\frac{f - g^*}{\|f - g^*\|_F}$ with norm 1.

We have

$$\left\| \frac{f - g^*}{\|f - g^*\|_F} - g \right\|_F = \frac{1}{\|f - g^*\|_F} \|f - g^* - g\,\|f - g^*\|_F\|_F .$$

Since $\overline{V}$ is a linear space, the vector $h := g^* + g\,\|f - g^*\|_F$ belongs to $\overline{V}$. Hence, $\|f - h\|_F \geq k$.

We thus have

$$\varepsilon = \|f - g^*\|_F \text{ and } \varepsilon \leq \|f - g^*\|_F \leq \frac{\varepsilon}{k},$$

from which

$$\left\| \frac{f - g^*}{\|f - g^*\|_F} - g \right\|_F = \frac{1}{\|f - g^*\|_F} \|f - h\|_F \geq \frac{\varepsilon}{\|f - g^*\|_F} \geq k.$$

This holds true for any $g \in \overline{V}$, which proves the claim. □

Finally, we may also characterize the dimension of a normed linear space through the concept of compactness. It holds:

THEOREM A1.5.– If the $\overline{S(f,1)}$ of a normed linear space $(V, \|.\|)$ is compact, then V is finite dimensional.

PROOF.– By contradiction.

Assume the sphere compact together with $\dim V = +\infty$.

Let φ_1 be a vector with norm 1 and let $span\,\{\varphi_1\}$ be the linear space generated by φ_1, $\dim\,(span\,\{\varphi_1\}) = 1$, $span\,\{\varphi_1\}$ is a closed space in V and $span\,\{\varphi_1\} \subset V$.

By the above Riesz Theorem we can find a vector with norm 1, say φ_2 such that $\|\varphi_2 - \varphi_1\| \geq \frac{1}{2}$.

The linear space $span\,\{\varphi_1, \varphi_2\} \subset V$ is closed in V as it has dimension 2. We define in this way a sequence of vectors $\varphi_1, \varphi_2, \varphi_3, \varphi_4, ..., \varphi_n, ...$, such that $\|\varphi_i - \varphi_j\| \geq \frac{1}{2}$, for any $i \neq j$, $i = 1, ..., n, ...$

Clearly, we cannot extract any convergent subsequence from this sequence $\left\{\varphi_n \in \overline{S(f,1)}, n \in \mathbb{N}\right\}$. □

THEOREM A1.6.– A normed linear space has a finite dimension if and only if its closed unit sphere is compact.

PROOF.– The condition is necessary since in a normed linear space a compact set is closed and bound. That the condition is sufficient comes from according to theorem A1.5. □

We now conclude that a normed linear space $(V, \|.\|)$ with finite dimension n is identical to $\mathbb{R}^n$ with respect to its topological properties. Denote $\{\varphi_1, ..., \varphi_n\}$, a basis of V.

It is verified that the isomorphism

$$\Psi_{\{\varphi_1,...,\varphi_n\}} : V_n \to \mathbb{R}^n, \ f \mapsto \Psi_{\{\varphi_1,...,\varphi_n\}}(f) := (\alpha_1, ..., \alpha_n)',$$

where, $(\alpha_1, ..., \alpha_n)'$ are the coordinates of the vector f in the basis $\{\varphi_1, ..., \varphi_n\}$, is a homeomorphism, i.e. is continuous with continuous inverse mapping $\Psi^{-1}_{\{\varphi_1,...,\varphi_n\}}$.

The proof proceeds by induction on n; see [KOL 74] and Bratti [BRA 98].

The conclusion of this set of results can now be drawn. If a "complex" object f (in an infinite dimensional space) has to be approximated, this should be achieved through a "simple" object φ. In fact an infinite dimensional linear space does not own convenient topological properties which makes an optimization problem tractable. On the contrary, such is the case of finite dimensional normed spaces.

Also, the choice of normed linear spaces to describe the setting of some optimization problems is justified by the fact that the object f may then be written as some linear combination of "simple" objects in a way such that error terms be amendable; the norm appears as a convenient tool to quantify the approximation of f by φ.

A1.2. Compatibility between the topological and the linear structure in linear spaces

For F a linear space equipped with a topology.

DEFINITION A1.6.– Say that the topology on F is compatible with the structure of linear space means that both mappings

$F^2 \to \mathbb{R}, (f, g) \mapsto f + g$ and $\mathbb{R} \times F \to \mathbb{R}, (\alpha, f) \mapsto \alpha f$ are continuous.

DEFINITION A1.7.– A set F is a real topological linear space, if and only if:

1) F is a linear space on $\mathbb{R}$;

2) F is a topological space;

3) the topology on F is compatible with the structure of linear space.

Continuity of the mapping $F^2 \to \mathbb{R}, (f,g) \mapsto f+g$ means that for any neighborhood $\mathcal{I}_{f+g}$ of $f+g$, there exist two neighborhoods $\mathcal{I}_f$ of f, and $\mathcal{I}_g$ of g, with

$$\mathcal{I}_f + \mathcal{I}_g \subset \mathcal{I}_{f+g}.$$

Continuity of the mapping $\mathbb{R} \times F \to \mathbb{R}, (\alpha, f) \mapsto \alpha f$ ensures that for any neighborhood $\mathcal{I}_{\alpha f}$ of αf there exist a real number $\beta(f)$ and a neighborhood of f, say $\mathcal{I}_f$, such that

$$\eta \mathcal{I}_f \subset \mathcal{I}_{\alpha f} \text{ if } |\eta - \alpha| < \beta(f),$$

where η is some real number. Recall that

$$A + B := \{f + g : f \in A, g \in B\},$$
$$\lambda A := \{\lambda f : f \in A, \lambda \in \mathbb{R}\}.$$

Topological linear spaces provide a convenient setting for the study of all topological properties of their elements. These follow properties of elements similar or close to the null vector.

Let $g \in F$. The translation mapping $t_g : F \to F, f \mapsto t_g(f) := f + g$ is one to one in F. Both t_g and t_g^{-1} are continuous mappings; t_g is an homeomorphism from F onto F.

The same holds for the homotheties with constant $\alpha \neq 0$ defined through $\mathbf{O}_\alpha : \mathbb{R} \times F \to F, f \mapsto \mathbf{O}_\alpha(f) := \alpha f$.

For any $f \in F$, all neighborhoods of f are obtained by translating the neighbors of the null vector; for $\mathcal{I}_0$ a generic such neighborhood, $\mathcal{I}_f$ is obtained through

$$\mathcal{I}_f = \mathcal{I}_0 + f.$$

A special case of a topological linear space is a normed linear space. In this case, the continuity of the sum $F^2 \to \mathbb{R}, (f,g) \mapsto f+g$ follows on from the triangle inequality

$$\|(f+g) - (h+i)\|_F \leq \|f - h\|_F + \|h - i\|_F$$

and the continuity of the mapping $\mathbb{R} \times F \to \mathbb{R}, (\alpha, f) \mapsto \alpha f$ follows on from

$$\lambda f - \lambda_0 f_0 = (\lambda - \lambda_0)(f - f_0) + (\lambda - \lambda_0) f_0 + (f - f_0) \lambda_0.$$

Extensions and deepening can be found for example in [BRY 08], or in Kolmogorov and Fomine [KOL 81].

A1.3. A characterization of the sup norm of a continuous function defined on a compact set

We briefly recall the following notation from Chapter 5.

We denote X a generic compact set in $\mathbb{R}^m$ and for any f in $\mathcal{C}^{(0)}(X,\mathbb{R})$ the critical set of f

$$\mathbf{Cr}(f) := \left\{y \in X : |f(y)| = \|f\|_{X,\infty}\right\}.$$

Next, let V be a finite dimensional subspace in $\mathcal{C}^{(0)}(X,\mathbb{R})$ generated by a basis $\{\varphi_1, ..., \varphi_k\}$.

Let

$$\mathbf{K} := \left\{(f(y)\varphi_1(y), ..., f(y)\varphi_k(y))', y \in \mathbf{Cr}(f)\right\} \subset \mathbb{R}^k$$

and

$$\widetilde{\mathbf{K}} := conv\{\psi_1, ..., \psi_r\text{: } \psi_i \in \mathbf{K}, i = 1, ..., r, r \leq k+1\}$$

It holds

THEOREM A1.7.– With the above notation

$$\|f+\varphi\|_{X,\infty} \geq \|f\|_{X,\infty}$$

for all φ in V, if and only if, $\underline{\mathbf{0}} := \left(0^1, ..., 0^k\right) \in \widetilde{\mathbf{K}}$.

PROOF (Rivlin [RIV 74, p. 61].– Since X is a compact set, it holds that $\mathbf{Cr}(f) \neq \phi$. Assume that $\underline{\mathbf{0}} \in \widetilde{\mathbf{K}}$; we prove that $\|f+\varphi\|_{X,\infty} \geq \|f\|_{X,\infty}$ for any φ in V. Since $\underline{\mathbf{0}} \in \widetilde{\mathbf{K}}$, there exists a linear convex combination of r vectors in $\mathbf{K}$ which equals $\underline{\mathbf{0}}$. Hence, there exist $\alpha_i \geq 0$, $1 \leq i \leq r$ with $1 = \sum_{i=1}^r \alpha_i$, such that $\underline{\mathbf{0}} = \sum_{i=1}^r \alpha_i (f(y_i)\varphi_1(y_i), ..., f(y_i)\varphi_k(y_i))'$. This yields the system $0 = \sum_{i=1}^r \alpha_i f(y_i)\varphi_j(y_i)$ for $j = 1, ..., k$. It follows that for any $\varphi \in V$, it holds that $0 = \sum_{i=1}^r \alpha_i f(y_i)\varphi(y_i)$. Therefore, $0 = \sum_{i=1}^r \alpha_i f(y_i)\varphi(y_i) = \sum_{i=1}^r \alpha_i f(y_i)\varphi(y_i)$. Hence, there exists some index i (depending on φ) with $f(y_i)\varphi(y_i) \geq 0$. Otherwise with $\alpha_i \geq 0$, and $1 = \sum_{i=1}^r \alpha_i$, it follows that $0 = \sum_{i=1}^r \alpha_i f(y_i)\varphi(y_i) < 0$, a contradiction. Let $\varphi \in V$ and i such that $f(y_i)\varphi(y_i) \geq 0$. Since $y_i \in \mathbf{Cr}(f)$, it holds

$$f(y_i)f(y_i) = \|f\|_{X,\infty}^2$$

since $y_i \in \mathbf{Cr}(f)$. Hence,

$$\begin{aligned}\| f+\varphi\|_{X,\infty}^2 &\geq | f(y_i)+\varphi(y_i)|^2 \\ &= (f(y_i)+\varphi(y_i))(f(y_i)+\varphi(y_i)) \\ &\geq \| f\|_{X,\infty}^2+\varphi^2(y_i)+2\sum_{i=1}^{r}\alpha_i f(y_i)\varphi(y_i) \\ &\geq \| f\|_{X,\infty}^2 .\end{aligned}$$

The last inequality is implied by $f(y_i)\varphi(y_i)\geq 0$.

We now turn to the reciprocal statement. Assume that $\|f+\varphi\|_{X,\infty}\geq \|f\|_{X,\infty}$ for all φ in V. Two cases may occur:

1) if f has norm zero, then $\|f\|_{X,\infty} = 0$ implies $f = \underline{\mathbf{0}}$ and therefore, $(f(y)\varphi_1(y),\ldots,f(y)\varphi_k(y))' = (\underline{\mathbf{0}},\ldots,\underline{\mathbf{0}})' = \underline{\mathbf{0}} \in \mathbf{K}$, which yields $\underline{\mathbf{0}} \in \widetilde{\mathbf{K}}$. Indeed since $\widetilde{\mathbf{K}}$ is the convex hull of $\mathbf{K}$, it holds $\mathbf{K} \subseteq \widetilde{\mathbf{K}}$;

2) if the norm of f is not zero, then $f/\| f\|_{X,\infty}$ has norm 1. Assume, therefore, that $\| f\|_{X,\infty}=1$. By contradiction, we prove that also in this case $\underline{\mathbf{0}} \in \widetilde{\mathbf{K}}$.

Assume that $\underline{\underline{\mathbf{0}}} \notin \widetilde{\mathbf{K}}$. Since, $\widetilde{\mathbf{K}}$ is convex, there exists an hyperplane which separates $\underline{\mathbf{0}}$ from $\widetilde{\mathbf{K}}$. Let $(c_0,c_1,\ldots,c_k)$ in $\mathbb{R}^{k+1}$ with $c_0\neq 0$, such that the half-space

$$\mathcal{S}^+ := \left\{(z_1,\ldots,z_k)\in\mathbb{R}^k : c_0+c_1z_{1+}\ldots+c_kz_k\geq 0 \text{ and } c_0=-\gamma<0\right\}$$

contains $\widetilde{\mathbf{K}}$.

Hence,

$$c_0+c_1z_{1+}\ldots+c_kz_k\geq 0 \Rightarrow c_1z_{1+}\ldots+c_kz_k\geq -c_0 := \gamma.$$

Since

$$(f(y)\varphi_1(y),\ldots,\ f(y)\varphi_k(y))' \in \mathbf{K} \subset \widetilde{\mathbf{K}}$$

it follows that for any $y\in\mathbf{Cr}(f)$, using

$$c_1z_{1+}\ldots+c_kz_k\geq -c_0 := \gamma,$$

and choosing

$$(z_1, ..., z_k) = (f(y)\varphi_1(y), ..., f(y)\varphi_k(y))'$$

it holds

$$\sum_{j=1}^{k} c_j f(y)\varphi_j(y) \geq \gamma > 0.$$

With $\varphi_0(y) := \sum_{j=1}^{k} c_j\varphi_j(y)$ in the span of the φ_j's. The last inequality is,

$$\sum_{j=1}^{k} c_j f(y)\varphi_j(y) = f(y)\sum_{j=1}^{k} c_j\varphi_j(y) \qquad \text{[A1.2]}$$

$$= f(y)\varphi_0(y) \geq \gamma > 0$$

for any $y \in \mathbf{Cr}(f)$.

Now, there exists an open set $\mathbf{A} \subset \mathbb{R}^m$ with $\mathbf{Cr}(f) \subset \mathbf{A}_1 := X \cap \mathbf{A}$ (where X is the domain of f) and it holds

$$\begin{cases} f(y)\varphi_0(y) \geq \frac{\gamma}{2} \text{ for all } y \text{ in } \mathbf{A}_1 \\ |f(y)| \leq 1 - \delta \text{ with } \delta > 0\text{, when } y \text{ is in the closed set } X\setminus\mathbf{A}_1. \end{cases}$$

As $\mathbf{Cr}(f)$ is a compact subset in $\mathbb{R}^m$ there exists such an open set $\mathbf{A}_1$ which contains $\mathbf{Cr}(f)$. Now, by [A1.2]

$$f(y)\varphi_0(y) \geq \gamma > \frac{\gamma}{2} > 0$$

it holds that on $\mathbf{A}_1$

$$f(y)\varphi_0(y) \geq \frac{\gamma}{2}$$

holds. On the contrary, it holds

$$|f(y)| \leq 1 - \delta \text{ with } \delta > 0$$

on the closed set $X\setminus\mathbf{A}_1$ since $|f|$ assumes its maximal values on each of the points $y \in \mathbf{Cr}(f)$ and only on those points; recall that by hypothesis $\|f\|_\infty := \max_{x\in X}|f(x)| = 1$. Let ε such that

$$0 < \varepsilon < \frac{\delta}{\|\varphi_0\|_{X,\infty}}$$

i.e.

$$\varepsilon \| \varphi_0 \|_{X,\infty} < \delta.$$

Such a positive ε exists as the system $\{\varphi_1, ..., \varphi_k\}$ is the basis, which implicates that

$$\varphi_0(y) := \sum_{j=1}^{k} c_i \varphi_j(y) = \mathbf{0}$$

cannot hold, since

$$(c_1, ..., c_k) \neq (0, ..., 0). \qquad [A1.3]$$

[A1.3] holds since otherwise, by

$$c_1 z_{1+} ... + c_k z_k \geq - c_0 := \gamma$$

$0 \geq \gamma$ would hold, a contradiction with $\gamma > 0$. This yields $\| \varphi_0 \|_{X,\infty} \neq 0$ and therefore $\delta / \| \varphi_0 \|_{X,\infty}$ is defined.

We now prove that it holds true:

a) for $y \in X \setminus \mathbf{A}_1$

$$\max | f(y) - \varepsilon \varphi_0(y) | < 1$$

b) for $y \in \mathbf{A}_1$

$$\begin{aligned} | f(y) - \varphi_0(y) |^2 &= | f(y) |^2 + \varepsilon^2 |\varphi_0(y)|^2 - 2\varepsilon f(y) \varphi_0(y) \\ &\leq 1 + \varepsilon^2 \|\varphi_0\|^2_{\infty, \mathbf{A}_1} - \varepsilon\gamma \end{aligned}$$

For case a), assume that $y \in X \setminus \mathbf{A}_1$ and prove that

$$\| f - \varepsilon \varphi_0 \|_{\infty, X \setminus \mathbf{A}_1} := \max_{X \setminus \mathbf{A}_1} | f(y) - \varepsilon \varphi_0(y) | < 1.$$

For case b), for any $y \in X \setminus \mathbf{A}_1$ it holds $\| f \|_{\infty, X \setminus \mathbf{A}} \leq 1 - \delta$. Indeed from

$$| f(y) | \leq 1 - \delta$$

it follows that

$$\|f\|_{\infty,\, X\setminus \mathbf{A}_1} := \max_{y\in X\setminus \mathbf{A}_1} |f(y)| \leq 1-\delta.$$

Along with the fact that $\varepsilon \| \varphi_0\|_{X,\infty} < \delta$, this implicates

$$\begin{aligned}\|f(y)-\varepsilon\varphi_0\|_{\infty,X\setminus \mathbf{A}_1} &\leq \|f\|_{\infty,X\setminus \mathbf{A}_1} + \|\varepsilon\varphi_0\|_{\infty,X\setminus \mathbf{A}_1}\\ &\leq 1-\delta+\|\varepsilon\varphi_0\|_{\infty,\, X\setminus \mathbf{A}_1}\\ &< 1-\delta+\delta = 1.\end{aligned}$$

Hence,

$$\|f-\ \varepsilon\varphi_0\|_{\infty,\, X\setminus \mathbf{A}_1} < 1$$

and, with $\varphi := -\ \varepsilon\varphi_0$,

$$\|f+\varphi\|_{\infty,\, X\setminus \mathbf{A}_1} < 1.$$

We will now prove the second case, for all $y \in \mathbf{A}_1$ it holds

$$\begin{aligned}&|\ f(y)-\ \varepsilon\varphi_0(y)|^2\\ &=|\ f(y)|^2+\varepsilon^2|\varphi_0(y)|^2-2\varepsilon f(y)\varphi_0(y)\\ &\leq 1+\varepsilon^2\|\varphi_0\|^2_{\infty,\mathbf{A}_1}-\varepsilon\gamma.\end{aligned}$$

Clearly,

$$\begin{aligned}|\ f(y)-\ \varepsilon\varphi_0(y)|^2 &= |\ f(y)|^2-2\varepsilon f(y)\varphi_0(y)+\varepsilon^2|\varphi_0(y)|^2\\ &=1-2\varepsilon f(y)\varphi_0(y)+\varepsilon^2\|\varphi_0\|^2_{\infty,\mathbf{A}_1}\\ &<1-\varepsilon\gamma+\varepsilon^2\|\varphi_0\|^2_{\infty,\mathbf{A}_1}\end{aligned}$$

where we used the fact that in $\mathbf{A}_1$ it holds $f(y) = 1$. Now, after mathematical treament, we find that,

$$|\ f(y)|^2+\varepsilon^2\|\varphi_0\|^2_{\infty,\mathbf{A}_1}-2\varepsilon f(y)\varphi_0(y)\leq 1+\varepsilon^2\|\varphi_0\|^2_{\infty,\mathbf{A}_1}-\varepsilon\gamma$$

if

$$f(y)\varphi_0(y)\geq\frac{\gamma}{2},$$

which holds true due to [A1.2]. Also, it holds that

$$| f(y)|^2 + \varepsilon^2 |\varphi_0(y)|^2 - 2\varepsilon f(y)\varphi_0(y) \leq 1 + \varepsilon^2 \|\varphi_0\|^2_{\infty,\mathbf{A}_1} - \varepsilon\gamma.$$

Hence,

$$| f(y) - \varepsilon\varphi_0(y)|^2 < 1 - \varepsilon\gamma + \varepsilon^2 \|\varphi_0\|^2_{\infty,\mathbf{A}_1}$$

which proves the claim.

We also have

$$\varepsilon^2 \|\varphi_0\|^2_{\infty,\mathbf{A}_1} - \varepsilon\gamma < 0.$$

Clearly,

$$\varepsilon^2 \|\varphi_0\|^2_{\infty,\mathbf{A}_1} - \varepsilon\gamma < 0 \text{ iff } \varepsilon^2 \|\varphi_0\|^2_{\infty,\mathbf{A}_1} < \varepsilon\gamma \text{ iff } \|\varphi_0\|^2_{\infty,\mathbf{A}_1} < \frac{\gamma}{\varepsilon}$$

holds if

$$\varepsilon < \frac{\gamma}{\|\varphi_0\|^2_{\infty,\mathbf{A}_1}}.$$

Now, this ultimate condition holds true since $\|\varphi_0\|^2_{\infty,\mathbf{A}_1} \neq 0$ and since ε is any positive number. For ε small enough ($\varepsilon < \gamma / \|\varphi_0\|^2_{\infty,\mathbf{A}_1}$), on all $\mathbf{A}_1$ it holds that

$$| f(y) - \varepsilon\varphi_0(y)|^2 < 1 - \varepsilon\gamma + \varepsilon^2 \|\varphi_0\|^2_{\infty,\mathbf{A}_1} < 1$$

and therefore,

$$| f(y) - \varepsilon\varphi_0(y)|^2 < 1 \text{ implies } | f(y) - \varepsilon\varphi_0(y)| < 1 \text{ for all } y \in \mathbf{A}_1.$$

Consider $\varphi := -\varepsilon\,\varphi_0$; it then holds

$$| f(y) + \varphi(y)| < 1 \text{ for all } y \in \mathbf{A}_1$$

hence

$$\max_{y\in\mathbf{A}_1} | f(y) + \varphi(y)| < 1$$

i.e.

$$\|f + \varphi\|_{\infty,A_1} < 1.$$

It follows that either on $\mathbf{A}_1$ or on $X \diagdown \mathbf{A}_1$ we obtain

$$\|f + \varphi\|_\infty < 1.$$

This contradicts the hypothesis stating that

$$\|f + \varphi\|_\infty \geq \|f\|_\infty, \text{ for all } \varphi \text{ in } span\{\varphi_1, ..., \varphi_k\}.$$

Indeed $\|f\|_\infty = 1$, which implies

$$\|f + \varphi\|_\infty \geq 1$$

and therefore,

$$\|f + \varphi\|_\infty \geq \|f\|_\infty.$$

It follows that $\underline{\mathbf{0}} \in \widetilde{\mathbf{K}}$.

Finally, since $\widetilde{\mathbf{K}}$ has r elements, according to the Carathéodory theorem, $r \leq k+1$. $\square$

Appendix 2

Chebyshev Polynomials

The Chebyshev polynomial of first kind of order n is

$$T_n : [-1;1] \to \mathbb{R}, x \mapsto T_n(x) := \cos(n(\arccos x)).$$

This is a polynomial with degree n; with $x = \cos\theta$,

$$T_n(x) = T_n(\cos\theta) = \cos(n(\arccos\cos\theta)) = \cos(n\theta).$$

From

$$\cos(n\theta) + i\sin(n\theta) = (\cos\theta + i\sin\theta)^n = \sum_{j=0}^{n} \binom{n}{j} i^j (\sin\theta)^j (\cos\theta)^{n-j},$$

$$\cos(n\theta) = \sum_{k=0}^{\left[\frac{n}{2}\right]} \binom{n}{2k} (-1)^k \left(1-\cos^2\theta\right)^k (\cos\theta)^{n-2k}.$$

Finally,

$$T_n(x) = \sum_{k=0}^{\left[\frac{n}{2}\right]} \binom{n}{2k} (-1)^k \left(1-x^2\right)^k x^{n-2k} = \sum_{k=0}^{\left[\frac{n}{2}\right]} \binom{n}{2k} \left(x^2-1\right)^k x^{n-2k}.$$

Since $\left(x^2-1\right)^k x^{n-2k}$ has degree n, it follows that the degree of T_n is less than or equal to n. To prove that this degree is n see that

$$T_n(x) = \sum_{k=0}^{\left[\frac{n}{2}\right]} \binom{n}{2k} \left(x^n - kx^{n-1} + \ldots\right) = x^n \sum_{k=0}^{\left[\frac{n}{2}\right]} \binom{n}{2k} - x^{n-1} \sum_{k=0}^{\left[\frac{n}{2}\right]} \binom{n}{2k} k + \ldots.$$

The coefficient of x^n is

$$\sum_{k=0}^{\left[\frac{n}{2}\right]} \binom{n}{2k} = 2^{n-1} \neq 0.$$

Hence, $\deg T_n = n$.

The expression $T_n := \cos\left(n\left(\arccos x\right)\right)$ provides the formula for T_n on $[-1,1]$ only. To obtain its expression on the real line, observe that there exists a unique polynomial P_n such that $P_n\left(\cos\theta\right) = \cos\left(n\theta\right)$, for all $\theta \in \mathbb{R}$. This coincides with $T_n := \cos\left(n\left(\arccos x\right)\right)$ on $[-1;1]$.

Furthermore, starting from this expression we find that T_n satisfies the equation, valid for all $n \geq 1$,

$$\begin{cases} T_{n+1}\left(x\right) = 2xT_n\left(x\right) - T_{n-1}\left(x\right) \\ T_0\left(x\right) = 1, T_1\left(x\right) = x \end{cases} \quad \text{[A2.1]}$$

For $x \in (-1;1)$ set $x = \cos\theta$ and obtain $T_n\left(x\right) = \cos\left(n\left(\arccos x\right)\right)$. Now,

$$\cos\left(n\theta\right) + \cos\left(\left(n+2\right)\theta\right) = 2\cos\theta\cos\left(\left(n+1\right)\theta\right)$$

holds true for all n.

Hence,

$$P_n\left(n\theta\right) + P_{n+2}\left(\left(n+2\right)\theta\right) = 2\cos\theta P_{n+1}\left(\left(n+1\right)\theta\right).$$

Change $\cos\theta$ into x to obtain

$$P_{n+1}\left(x\right) = 2xP_n\left(x\right) - P_{n-1}\left(x\right).$$

Furthermore, the solution to this equation is unique. Consider the variable in $[1;\infty)$ instead of $[-1;1]$; change θ into y through $y = \cosh\theta$, per $x > 1$, yielding

$$T_n\left(\cosh\theta\right) = \cosh\left(n\theta\right)$$

from which we get the definition of the Chebyshev polynomials on $[1;\infty)$. The definition on $(-\infty;1]$ is deduced from $T_n\left(x\right) = T_n\left(-x\right)$. With some calculationss, it holds for all $x \in \mathbb{R}$:

$$T_n\left(x\right) = \frac{1}{2}\left(\left(x+\sqrt{x^2-1}\right)^n + \left(x-\sqrt{x^2-1}\right)^n\right).$$

If $x \in [a;b]$ the change of variable $y = \frac{2x-a-b}{b-a}$ brings back to $[-1,1]$.

Summing up, we have

$$\begin{aligned} T_n(x) &= 2^{n-1} \prod_{j=1}^{n} \left(x - \cos\left(\frac{\pi(2j+1)}{2n} \right) \right) \\ &= \frac{1}{2} \left\{ \left(x + \sqrt{x^2-1} \right)^n + \left(x - \sqrt{x^2-1} \right)^n \right\} \\ &= \begin{cases} \cos(n \arccos x) & \text{for } |x| \leq 1 \\ (-1)^{n\, sign(x)} \cosh\left(n \cosh^{-1} |x| \right) & \text{for } |x| \geq 1 \end{cases} \end{aligned}$$

By the d'Alembert-Gauss theorem it follows that the equation

$$T_n(x) = 0,$$

has n roots.

These are given by the equation

$$\cos(n(\arccos x_k)) = 0$$

i.e.

$$n(\arccos x_k) = \frac{\pi}{2} + k\pi = \frac{(2k+1)\pi}{2}$$

i.e.

$$\arccos x_k = \frac{(2k+1)\pi}{2n}$$

i.e.

$$x_k = \cos\left(\frac{(2k-1)\pi}{2n} \right), k = 1, ..., n.$$

Hence, $x_k \in (-1;1)$. Furthermore, it can be proved that the number of roots of T_n in some interval $[x; x+\delta] \subset [-1;1]$ is of order $\frac{n\delta}{\pi\sqrt{1-x^2}}$ as $n \to \infty$. Hence, the roots of T_n tend to densify at the extremities of $[-1;1]$.

All roots are simple since the function cos is strictly decreasing on $[0;\pi]$.

Also,

$$T_n(x) = 2^{n-1} \prod_{j=1}^{n} \left(x - \cos \left(\frac{(2j+1)\pi}{2} \right) \right).$$

It holds for $x \in (-1; 1)$,

$$\frac{dT_n(x)}{dx} = \frac{d(\cos(n(\arccos x)))}{dx} = \frac{n \sin(n(\arccos x))}{\sqrt{1-x^2}}.$$

The minimal and maximal values of T_n are respectively

$$x_h^* := \cos \frac{h}{n}\pi, \; h = 0, ..., n, \;\; \text{or, } x_h^* := -\cos\left(\pi - \frac{h}{n}\pi\right), \; h = 0, ..., n\,.$$

They are pairwise distinct and

$$x_0^* = 1 > ... > x_h^* > ... > x_n^* = -1.$$

Furthermore,

$$|T_n(x_h^*)| = 1.$$

From x_h^* to x_{h+1}^*, T_n oscillates between -1 and 1. Indeed,

$$\begin{aligned} T_n(x_h^*) &= T_n\left(\cos \frac{h}{n}\pi\right) = \cos\left(n\left(\arccos\left(\cos \frac{h}{n}\pi\right)\right)\right) \\ &= \cos\left(n \frac{h}{n}\pi\right) = \cos(h\pi) = (-1)^h, h = 0, ..., n. \end{aligned}$$

and on the minimum/maximum points x_h^*

$$|T_n(x_h^*)| = \left|(-1)^h\right| = 1, h = 0, ..., n.$$

Such a set $\left\{x_h^* = \cos \frac{h}{n}\pi, \; h = 0, ..., n, \right\}$ is called alternating for T_n.

Appendix 3

Some Useful Inequalities for Polynomials

A3.1. Bounds on the derivatives of polynomials on compact sets

We first consider inequalities pertaining to the derivatives of polynomials; see Szegö [SZE 75].

For a given function

$$f : [a.b] \to \mathbb{R},$$

with $f \in \mathcal{C}^{(\infty)}([a,b])$ and for which

$$|f(x)| \leq L, \; x \in [a,b],$$

the question is: find minimal constants $M(i)$ such that

$$\left| \frac{d^i f(x)}{dx^i} \right| \leq M(i), \; x \in [a,b].$$

For some special functions the answer to this question is simple. Those functions are those which coincide with their derivative up to a constant. In this case, $M(i)$ is related to L.

For example, the functions

$$e^{mx}, \;\; \sin(mx), \;\; \cos(mx), \; m \in \mathbb{R}$$

are of this kind. Indeed,

$$|e^x| \leq L \Rightarrow \left|\frac{d^i e^{mx}}{dx^i}\right| = m^i |e^x| \leq m^i L,$$

and since

$$\frac{d^i \sin x}{dx^i} = \sin\left(x + \frac{i\pi}{2}\right), \quad \frac{d^i \cos x}{dx^i} = \cos\left(x + \frac{i\pi}{2}\right),$$

it follows that

$$|\sin(mx)| \leq L \Rightarrow \left|\frac{d^i \sin(mx)}{dx^i}\right| = m^i \left|\sin(mx) + \frac{i\pi}{2}\right| \leq m^i \left(L + \frac{i\pi}{2}\right),$$

$$|\cos(mx)| \leq L \Rightarrow \left|\frac{d^i \cos(mx)}{dx^i}\right| = m^i \left|\cos(mx) + \frac{i\pi}{2}\right| \leq m^i \left(L + \frac{i\pi}{2}\right).$$

Henceforth, for those functions the upper bound on the function itself induces an upper bound on its derivatives.

More generally, since the derivation is a linear operator which holds for the functions

$$e^{mx}, \ \sin(mx), \ \cos(mx), \ m \in \mathbb{R},$$

also holds for any finite linear combination of those. Such as,

$$\mathbf{T}_n(x) := \frac{1}{2}a_0 + \sum_{j=1}^{n} \left(a_j \cos(jx) + b_j \sin(jx)\right),$$

which are the trigonometric polynomials of order $n \geq 1$, which bear a strong link with $\mathcal{P}_{2n}$, the class of all polynomials with degree less or equal $2n$, since with any trigonometric polynomial with degree n on $\mathbb{R}$ we can associate a corresponding polynomial in $\mathcal{P}_{2n}$, as proved in the next Lemma. This fact will be used to provide bounds for the derivatives of a function f in $\mathcal{P}$. Instead of considering polynomials $P_{2n} \in \mathcal{P}_{2n}$, we will consider linear combinations of the form $\mathbf{T}_n(x)$.

It holds

LEMMA A3.1 (Vitali).– Let

$$\mathbf{T}_n(x) := \frac{1}{2}a_0 + \sum_{j=1}^{n} (a_j \cos(jx) + b_j \sin(jx))$$

be a non-null trigonometric polynomial of order $n \geq 1$ (or so-called degree, meaning that a_n and b_n do not both equal 0), with real coefficients. Then $\mathbf{T}_n$ has at most $2n$ roots.

PROOF.– Consider the change of variable

$$x \mapsto t := e^{ix}$$

with $i := \sqrt{-1}$. By Euler formula, it holds

$$\cos jx = \frac{t^j + t^{-j}}{2}, \quad \sin jx = \frac{t^j - t^{-j}}{2i}.$$

Therefore,

$$\begin{aligned}
\mathbf{T}_n(x) &:= \frac{1}{2}a_0 + \sum_{j=1}^{n} (a_j \cos(jx) + b_j \sin(jx)) \\
&= \frac{1}{2}a_0 + \sum_{j=1}^{n} \left(a_j \frac{t^j + t^{-j}}{2} - ib_j \frac{t^j - t^{-j}}{2i}\right) \\
&= \frac{1}{2}a_0 + \frac{1}{2}\sum_{j=1}^{n} \left(a_j \left(t^j + t^{-j}\right) + ib_j \frac{t^j - t^{-j}}{i^2}\right) \\
&= \frac{1}{2}a_0 + \frac{1}{2}\sum_{j=1}^{n} \left(a_j \left(t^j + \frac{1}{t^j}\right) - ib_j \left(t^j - \frac{1}{t^j}\right)\right) \\
&= \frac{1}{2}a_0 + \frac{1}{2}\sum_{j=1}^{n} \left(a_j \left(\frac{t^{2j} + 1}{t^j}\right) \frac{t^{n-j}}{t^{n-j}} - ib_j \left(\frac{t^{2j} - 1}{t^j}\right) \frac{t^{n-j}}{t^{n-j}}\right) \\
&= \frac{1}{2}a_0 + \frac{1}{2}\sum_{j=1}^{n} \left(a_j \left(t^{n+j} + 1\right) \frac{1}{t^{2n}} - ib_j \left(t^{n+j} - 1\right) \frac{1}{t^{2n}}\right) \\
&= \frac{1}{t^{2n}} \left(\frac{1}{2}a_0 + \frac{1}{2}\sum_{j=1}^{n} \left(a_j \left(t^{n+j} + 1\right) - ib_j \left(t^{n+j} - 1\right)\right)\right) \\
&=: \frac{P_{2n}}{t^{2n}}.
\end{aligned}$$

We have proved that P_{2n} is a polynomial with real coefficients and degree $2n$ in the variable t. Since $t := e^{ix}$ uses the principal value of the logarithm function on $\mathbb{C}$, namely $x = \ln t + 2h\pi$, $h \in \mathbb{Z}$.

Since with all distinct values of x modulo 2π, we associate distinct values for t, and since P_{2n} has at most $2n$ roots, with each of them counted with its own multiplicity, the claim of the lemma is proved. □

We will now consider some technical results pertaining to the combination of polynomials, and the resulting properties of their derivatives. We first state a well-known result about the roots of polynomials.

PROPOSITION A3.1.– Let $f(x) \in \mathbb{k}[x]$ (the class of all polynomials with coefficients in the field $\mathbb{k}$). Then, α is a root of f with order larger than or equal to 2 if and only if $f'(x) = 0$.

PROOF.– Let α be a root of f with order larger than or equal to 2. Then

$$f(x) = (x - \alpha)^2 Q(x).$$

Derivation yields

$$\begin{aligned} f'(x) &= 2(x - \alpha) Q(x) + (x - \alpha)^2 Q'(x) \\ &= (x - \alpha)\left(2Q(x) + (x - \alpha)^2 Q'(x)\right). \end{aligned}$$

Hence,

$$f'(\alpha) = (\alpha - \alpha)\left(2Q(\alpha) + (\alpha - \alpha)^2 Q'(x)\right) = 0.$$

Assume that $f'(\alpha) = 0$.

Since α is a root of f, it holds $f(x) = (x - \alpha) Q(x)$. Derivating

$$f'(x) = Q(x) + (x - \alpha) Q'(x).$$

Since $f'(\alpha) = 0$,

$$f'(\alpha) = 0 = Q(\alpha) + (\alpha - \alpha) Q'(\alpha).$$

Hence, $Q(\alpha) = 0$. and α is a root of Q. Therefore,

$$Q(x) = (x - \alpha) S(x).$$

Since $f(x) = (x-\alpha) Q(x)$,

$$f(x) = (x-\alpha) Q(x) = (x-\alpha)^2 S(x).$$

Hence, α is a root of f with multiplicity larger or equal 2. □

LMMA A3.2 (Vitali).– Let $\mathbf{T}_n$ be a trigonometric polynomial with real coefficients and order $n \geq 1$. If its derivative $\mathbf{T}'_n$ satisfies

$$max_x |\mathbf{T}'_n(x)| = nL$$

where, $L > 0$ and $n \in \mathbb{N} \setminus \{0\}$, then there exists c such that the function

$$x \to L\sin(nx+c)$$

be such that $\frac{d}{dx}(L\sin(nx+c) - \mathbf{T}_n(x))$ has at least a root with multiplicity 2.

PROOF.– By the preceding Proposition A3.1 it is enough to prove that the function

$$x \to \frac{d^2}{dx^2}(L\sin(nx+c) - \mathbf{T}_n(x))$$

assumes value 0 at some point.

Let η be some point where the derivative of $\mathbf{T}_n$ assumes maximal absolute value, $\mathbf{T}'_n(\eta) = \epsilon nL$, with $\epsilon \in \{-1, 1\}$. It follows that $\mathbf{T}''_n(\eta) = 0$. Determine the constant c such that

$$\cos(n\eta + c) = \epsilon$$

(with $c = \pi h - n\eta, h \in \mathbb{Z}$).

If $c = \pi h - n\eta$, then $\sin(n\eta + \pi h - n\eta) = 0$. Hence, for this c it holds

$$\sin(n\eta + c) = 0$$

and therefore,

$$\mathbf{T}'_n(\eta) - \frac{dL\sin(n\eta+c)}{dx} = 0,$$

$$\mathbf{T}''_n(\eta) - \frac{d^2L\sin(n\eta+c)}{dx^2} = 0.$$

□

LEMMA A3.3 (Vitali).– Let $\mathbf{T}_n$ be a non-null trigonometric polynomial with real coefficients and order $n \geq 1$, and

$$max_{x\in[0,2\pi]} \left|\mathbf{T}_n\left(x\right)\right| = L_1.$$

Then, for any $L \geq L_1$, and any c the equation

$$\frac{d}{dx}\left(L\sin\left(nx+c\right) - \mathbf{T}_n\left(x\right)\right) = 0$$

has at least $2n$ roots.

PROOF.– We first consider $L_1 < L$. Consider the $2n+1$ points

$$x_1 := \frac{\pi}{2n} - \frac{c}{n}, ..., x_k := (2k-1)\frac{\pi}{2n} - \frac{c}{n}, ...,$$

$$x_n := (2n-1)\frac{\pi}{2n} - \frac{c}{n}, x_{2n+1} := x_1 + 2\pi.$$

Clearly,

$$f\left(x\right) := L\sin\left(nx+c\right) - \mathbf{T}_n\left(x\right).$$

Then,

$$f\left(x_1\right) = L\sin\left(n\left(\frac{\pi}{2n} - \frac{c}{n}\right) + c\right) - \mathbf{T}_n\left(\frac{\pi}{2n} - \frac{c}{n}\right) = L - \mathbf{T}_n\left(\frac{\pi}{2n} - \frac{c}{n}\right) > 0,$$

$$f\left(x_2\right) = L\sin\left(n\left(\frac{3\pi}{2n} - \frac{c}{n}\right) + c\right) - \mathbf{T}_n\left(\frac{3\pi}{2n} - \frac{c}{n}\right) = -L - \mathbf{T}_n\left(\frac{3\pi}{2n} - \frac{c}{n}\right) < 0,$$

since $max_{x\in[0,2\pi]} \left|\mathbf{T}_n\left(x\right)\right| = L_1 < L$. Hence, on those points the function $x \to f\left(x\right)$, alternates sign as L and $-L$. Hence, on any point inside the $2n$ intervals

$$]x_1, x_2[, ...,]x_{2n}, x_1 + 2\pi[$$

its derivative alternates sign. Therefore, f' has at least $2n-1$ distinct roots inside $]x_1, x_1+2\pi[$. Let $\xi_1, ..., \xi_{2n-1}$ be those ordered roots. From ξ_i to ξ_{i+1} the function f' changes sign. We see that whenever $f'\left(x_1\right) = 0$, then the claim of the Lemma holds.

Assume

$$f'\left(x_1\right) \neq 0.$$

By periodicity it then holds $f'(x_1 + 2\pi) \neq 0$. Since f' has only $2n - 1$ roots then inside $]x_1, \xi_1[$ the sign of f' is the sign of $f'(x_1)$. This sign is the opposite as the sign of f' on the intervals

$$]\xi_1, \xi_2[\,,]\xi_3, \xi_4[\,...,]\xi_{2n-1}, x_1 + 2\pi[\,.$$

Therefore, the sign of $f'(x_1)$ differs from the sign of $f'(x_1 + 2\pi)$, a contradiction. It follows that when $L_1 < L$, f' has, at least, $2n$ roots.

We now prove that the same holds when $L_1 = L$. By hypothesis,

$$\max_{x\in[0,2\pi]} |\mathbf{T}_n(x)| = L.$$

If

$$|\mathbf{T}_n(x)| \neq L,$$

the same argument used above at points

$$x = x_k := (2k-1)\frac{\pi}{2n} - \frac{c}{n}, k = 1, ..., 2n,$$

holds.

Assume, therefore, that for any x_k,

$$|\mathbf{T}_n(x_k)| = L.$$

Since, L is the maximal value of $\mathbf{T}_n$, and since the x_k's are interior points in $[0, 2\pi]$, the first order condition yields

$$\mathbf{T}_n'(x_k) = 0, k = 1, .., 2n.$$

Hence, the x_k s are roots of

$$\frac{d\left(L\sin(nx+c) - \mathbf{T}_n(x)\right)}{dx}.$$

Since, there are exactly $2n$ such x_k, the lemma is proved also in this case.

We consider the last case, assuming that for some of the x_ks,

$$|\mathbf{T}_n(x_k)| = L$$

and, for the others

$$|\mathbf{T}_n(x_k)| < L.$$

To get some intuition, assume for example that

$$\mathbf{T}_n(x_1) = L, ..., \mathbf{T}_n(x_k) = (-1)^{k-1} L,$$
$$|\mathbf{T}_n(x_{k+1})| \neq L, ..., |\mathbf{T}_n(x_{2n})| \neq L.$$

The points $x_1, ..., x_k$ are points of maximal value of $\mathbf{T}_n$, hence are roots of $\mathbf{T}'_n$ and also of

$$x \to \frac{d(L\sin(nx+c) - \mathbf{T}_n(x))}{dx}$$

which is written

$$\mathbf{T}_n(x_{k+r}) \neq (-1)^{k+r-1} L\sin(nx_{k+r}+c),$$
$$\mathbf{T}_n(x_{k+k'}) = (-1)^{k+k'-1} L\sin(nx_{k+k'}+c),$$

$r = 1, ..., k'-1$; $k' \leq 2n+1-k$, in some point interior to any of the intervals $]x_{k+r-1}, x_{k+r}[$. Then,

$$x \to \frac{d(L\sin(nx+c) - \mathbf{T}_n(x))}{dx},$$

has alternating signs and has at least $k'-1$ roots inside $]x_{k+r-1}, x_{k+r}[$ and one at point $x_{k+k'}$. Hence, the function

$$x \to \frac{d(L\sin(nx+c) - \mathbf{T}_n(x))}{dx}$$

has $k+k'$ roots in $]x_{k+r-1}, x_{k+r}[$. If $k+k' = 2n+1$, independently upon x_{2n+1}, we then have $2n$ roots and the lemma is proved. If $k+k' < 2n+1$, the $k+k'$ roots are all distinct points modulo 2π. The same argument holds beginning from $x_{k+k'}$. □

The preceding lemmas give some information on the "shape" of the trigonometric polynomials. It holds

PROPOSITION A3.2 (Vitali).– Let

$$\mathbf{T}_m(x) := \frac{1}{2}a_0 + \sum_{j=1}^{m} (a_j\cos(jx) + b_j\sin(jx)),$$

be a trigonometric polynomial with real coefficients and degree $m \geq 1$, such that $|a_m|+|b_m| \neq 0$ and, for any x, $|\mathbf{T}_m(x)| \leq L$ (with $L > 0$). If there exists some $n > 0$ and some point η, such that $|\mathbf{T}'_m(\eta)| = nL$, then either $\mathbf{T}_m(x) = L \sin(nx+c)$ or the order m satisfies $m > n$.

PROOF.– By hypothesis, $|\mathbf{T}'_m(\eta)| = nL$. Therefore, either $\mathbf{T}'_m$ has maximal value in η, or the maximal value of $\mathbf{T}'_m$ exceeds $\mathbf{T}'_m(\eta)$. If $\max_x |\mathbf{T}'_m(x)| > nL$ then, we may consider a real number λ such that $0 < \lambda < 1$ and $\max_x |\lambda \mathbf{T}'_m(x)| = nL$.

If $\max_x |\lambda \mathbf{T}'_m(x)| = nL$, by lemma A3.1 above, there exists some c such that

$$x \to \frac{d\left(L \sin(nx+c) - \lambda \mathbf{T}_m(x)\right)}{dx},$$

has a least a root of order 2.

By hypothesis, $|\mathbf{T}_m(x)| \leq L$ (with $L > 0$). Hence,

$$|\lambda \mathbf{T}_m(x)| \leq \lambda L < L.$$

By lemma A3.3 above the equation

$$\frac{d\left(L \sin(nx+c) - \lambda \mathbf{T}_m(x)\right)}{dx} = 0$$

has at least $2n$ roots and one of those has order 2. Hence, the function

$$x \to \frac{d\left(L \sin(nx+c) - \lambda \mathbf{T}_m(x)\right)}{dx},$$

has at least $2n+1$ roots.

The function

$$x \to L \sin(nx+c) - \lambda \mathbf{T}_m(x)$$

is not the null function; otherwise

$$\mathbf{T}_m(x) = \frac{L \sin(nx+c)}{\lambda},$$

and therefore,

$$\max_x \mathbf{T}_m(x) = \max_x \frac{L\ \sin(nx+c)}{\lambda} = \frac{L}{\lambda} > L\,,$$

a contradiction with $|\mathbf{T}_m(x)| \leq L$. Hence

$$x \to L\sin(nx+c) - \lambda\mathbf{T}_m(x) \neq 0.$$

By lemma A3.1, since there are at least $2n$ roots for $L\sin(nx+c) - \lambda\mathbf{T}_m(x) = 0$ it follows that the trigonometric polynomial

$$x \to L\sin(nx+c) - \lambda\mathbf{T}_m(x)$$

and the polynomials $\lambda\mathbf{T}_m$ and $\mathbf{T}_m$ have order m larger than n.

We consider the case when $\max_x |\mathbf{T}'_m(x)| = nL$. The function

$$x \to \frac{d\left(L\sin(nx+c) - \mathbf{T}_m(x)\right)}{dx}$$

has at least $2n+1$ roots. Since, we may also have

$$L\sin(nx+c) - \mathbf{T}_m(x) = 0,$$

it follows that either $\mathbf{T}_m(x) = L\sin(nx+c)$ or the order $m$of $\mathbf{T}_m$islargerthan n.□

Finally from the above Proposition A3.1 we infer the following corollary.

COROLLARY A3.1 (Bernstein).– If the absolute value of a trigonometric polynomial of order n with real coefficients does not exceed L, then the absolute value of its derivative of order k does not exceed $n^k L$.

PROOF.– By proposition A3.2, it holds

$$\mathbf{T}_n(x) = L\ \sin(nx+c).$$

Derivation yields

$$\mathbf{T}_n^{(k)}(x) = \frac{d^k\left(L\ \sin(nx+c)\right)}{dx^k} = L\, n^k\ \sin\left(nx + c + k\frac{\pi}{2}\right).$$

By hypothesis

$$|\mathbf{T}_n(x)| \leq L.$$

Hence,

$$\left|\mathbf{T}_n^{(k)}(x)\right| \leq L\, n^k.$$

THEOREM A3.1 (Bernstein).– Let P be a polynomial with degree n, and let $\max_{x\in[-1,1]} |P(x)| \leq L$. Then,

$$\left|\frac{dP(x)}{dx}\right| \leq \frac{nL}{\sqrt{(1-x^2)}} \text{ for any } x \in \left]-1,1\right[.$$

□

PROOF.– Through the change of variable $x = \cos\theta$, P becomes a trigonometric polynomial $\theta \to P(\cos\theta)$ of order n.

Derivation yields

$$\frac{dP(\cos\theta)}{d\theta} = \sin\theta \, \frac{dP(\cos\theta)}{d\theta}\frac{d\theta}{dx} = \sqrt{(1-x^2)}\,\frac{dP(x)}{dx}.$$

From

$$\max_{x\in[-1,1]} |P(x)| \leq L,$$

it follows that for all x

$$|P(x)| \leq L.$$

Hence, for all θ

$$|P(\cos\theta)| \leq L.$$

By theorem A3.1

$$\left|\frac{dP(\cos\theta)}{d\theta}\right| \leq nL,$$

and since

$$\left|\frac{dP(\cos\theta)}{d\theta}\right| = \left|\sqrt{(1-x^2)}\,\frac{dP(x)}{dx}\right|,$$

we get

$$\left|\frac{dP(x)}{dx}\right| \leq \frac{nL}{\sqrt{(1-x^2)}}.$$

□

The above theorem may be adapted to any interval $[a, b]$. It holds

PROPOSITION A3.3.– Let $P(x) = \sum_{i=0}^{n} a_i x^i$, $x \in [a, b]$ and let L be such that $\max_{x\in[a,b]} |P(x)| = L$. Then, for any $x \in]a, b[$, it holds

$$\left|\frac{dP(x)}{dx}\right| \leq \frac{nL}{\sqrt{(x-a)(b-x)}}.$$

PROOF.– The mapping

$$\varphi :]a, b[\ \rightarrow \ ]-1, 1[\ ; x \mapsto t = \varphi(x) := \frac{2x - a - b}{b - a},$$

transforms the polynomial $P(x)$ into the polynomial $Q(t)$ with degree n, with

$$\max_{t\in[-1,1]} |Q(t)| \leq L.$$

Applying the just above Theorem A3.1 to Q, we get for $t \in]-1, 1[$,

$$\left|\frac{dQ(t)}{dt}\right| \leq \frac{nL}{\sqrt{(1-t^2)}}.$$

Now

$$\frac{dP(t)}{dt} = \frac{dQ(\varphi(x))}{dx}\frac{d\varphi(x)}{dx} = \frac{2}{b-a}\frac{dQ(t)}{dt},$$

and

$$1 - t^2 = 1 - \left(\frac{2x - a - b}{b - a}\right)^2 = 4\frac{(x-a)(b-x)}{(b-a)^2}.$$

Therefore,

$$\left|\frac{dP(t)}{dt}\right| \leq \frac{nL}{\sqrt{(x-a)(b-x)}}.$$

□

The following result holds for derivatives with higher orders; it is due to Bernstein (see Vitali [VIT 51, pp. 461]).

THEOREM A3.2 (Bernstein).– Let $P(x) = \sum_{i=0}^{n} a_i x^i$, $x \in [-1,1]$ and let $\max_{x\in[-1,1]} |P(x)| \leq L$. Then, for any $x \in]-1,1[$, we have

$$\left|\frac{d^k P(x)}{dx^k}\right| \leq \frac{k^{\frac{k}{2}} n(n-1)\ldots(n-k+1) L}{(1-x^2)^{\frac{k}{2}}}.$$

PROOF.– Consider $x^* \in [-u,u]$ with $u \in]0,1]$. From Corollary A3.1 under the present hypotheses

$$\left|\frac{dP(x)}{dx}\right| \leq \frac{nL}{\sqrt{(1-x^2)}}, \text{ for any } x \in]-1,1[.$$

Therefore, for x^* in $[-u,u]$ and $u \in]0,1]$

$$\left|\frac{dP(x^*)}{dx}\right| \leq \frac{nL}{\sqrt{(1-x^{*2})}} \leq \frac{nL}{\sqrt{(1-u^2)}}.$$

Apply Proposition A3.1 to the function $x \to \frac{dP(x^*)}{dx}$ (with degree $n-1$), in the interval $[a,b]$, $a=-1$, $b=1$, and with the constant $\frac{nL}{\sqrt{(1-x^{*2})}}$. This yields

$$\left|\frac{d\left(\frac{dP(x)}{dx}\right)}{dx}\right| \leq \frac{(n-1)\frac{nL}{\sqrt{(1-x^{*2})}}}{\sqrt{(u+x)(u-x)}} = \frac{(n-1)\ n\ L}{\sqrt{(u^2-x^2)}\sqrt{(1-u^2)}}.$$

For fixed x in $]0,1]$ let $x_1 > 0$ be defined by

$$1-x_1^2 = x_1^2 - x^2 = \frac{1-x^2}{2}, x_1 > |x|$$

Then,

$$\left|\frac{d^2 P(x)}{dx^2}\right| \leq \frac{(n-1)\ n\ L}{\sqrt{(x_1^2-x^2)}\sqrt{(1-x_1^2)}} = \frac{(n-1)\ n\ L}{\sqrt{\frac{1-x^2}{2}}\sqrt{\frac{1-x^2}{2}}} = \frac{2\ (n-1)\ n\ L}{1-x^2}.$$

In the same way for fixed x and x_1 define x_2 such that

$$1-x_1^2 = x_1^2 - x_2^2 = x_2^2 - x^2 = \frac{1-x^2}{3},$$

for $-x_2 \leq x^{**} \leq x_2$, we deduce from

$$\left| \frac{d^2 P(x)}{dx^2} \right| \leq \frac{(n-1)\ n\ L}{\sqrt{(x_1^2 - x^2)}\sqrt{(1-x_1^2)}}$$

that

$$\left| \frac{d^2 P(x^{**})}{dx^2} \right| \leq \frac{(n-1)\ n\ L}{\sqrt{(x_1^2 - x^2)}\sqrt{(1-x_1^2)}}.$$

Apply Proposition A3.3 to the polynomial

$$x \to \frac{d^2 P(x)}{dx^2}$$

with degree $n-2$, we get

$$\left| \frac{d^3 P(x)}{dx^3} \right| \leq \frac{(n-2)(n-1)\ n\ L}{\sqrt{(x_2^2 - x^2)}\sqrt{(x_1^2 - x_2^2)}\sqrt{(1-x_1^2)}}.$$

Substituting

$$1 - x_1^2 = \frac{1-x^2}{3},\ x_1^2 - x_2^2 = \frac{1-x^2}{3},$$

we get

$$\left| \frac{d^3 P(x)}{dx^3} \right| \leq \frac{3^{\frac{3}{2}}(n-2)(n-1)\ n\ L}{(1-x_1^2)^{\frac{3}{2}}}.$$

Proceed in this way fixing x and determining positive numbers $x_1, ..., x_k$ such that

$$1 - x_1^2 = x_1^2 - x_2^2 = ... = x_{k-1}^2 - x^2 = \frac{1-x^2}{k}.$$

Apply Proposition A3.3 to the polynomial

$$x \to \frac{d^{k-1} P(x)}{dx^{k-1}},$$

with degree $n-k+1$ we get

$$\left| \frac{d^k P(x)}{dx^k} \right| \leq \frac{n(n-1)(n-2)...\ (n-k+1)\ L}{\sqrt{(x_{k-2}^2 - x_{k-1}^2)}\sqrt{(x_1^2 - x_2^2)}\sqrt{x_{k-1}^2 - x^2}\sqrt{(1-x_1^2)}}.$$

Finally, since $1 - x_1^2 = \frac{1-x^2}{k}$, $x_1^2 - x_2^2 == \frac{1-x^2}{k}$,...., $x_{k-1}^2 - x^2 = \frac{1-x^2}{k}$,

$$\left|\frac{d^k P(x)}{dx^k}\right| \leq \frac{k^{\frac{k}{2}} n(n-1)...(n-k+1)L}{(1-x^2)^{\frac{k}{2}}}.$$

□

The following result is a plain consequence of the above result.

COROLLARY A3.3.– Let P be a polynomial with degree n. Assume that

$$\max_{x\in[a,b]} |P(x)| \leq L.$$

Then, for any $x \in]a, b[$

$$\left|\frac{d^k P(x)}{dx^k}\right| \leq \left(\frac{k}{(b-x)(x-a)}\right)^{\frac{k}{2}} n(n-1)...(n-k+1)L.$$

PROOF.– Use the change of variable

$$\varphi :]a, b[\;\rightarrow\;]-1, 1[\;; x \mapsto t = \varphi(x) := \frac{2x-a-b}{b-a}.$$

□

Until now, we considered local bounds for the derivatives. We will consider uniform bounds now.

It holds

THEOREM A3.3 (Markov).– Let $P(x) := \sum_{j=0}^{n} a_j x^j$ be a polynomial with real coefficients such that $\sup_{x\in[a,b]} |P(x)| \leq W$, then for all x in (a, b) it holds

$$\left|P^{(i)}(x)\right| \leq \frac{n^2(n^2-1)...\left(n^2-(i-1)^2\right)}{(2i-1)!!}\left(\frac{2}{b-a}\right)^i W.$$

Furthermore, this upper bound is reached if x satisfies

$$P(x) = \pm W \cos\left(n \arccos \frac{2x-a-b}{b-a}\right). \quad \text{[A3.1]}$$

PROOF.– The difficult proof of this theorem has been simplified by Bernstein in an asymptotic setting. This latest proof will be presented later on.

We only prove that the maximal value

$$\frac{n^2\left(n^2-1\right)...\left(n^2-(i-1)^2\right)}{(2i-1)!!}\left(\frac{2}{b-a}\right)^i W,$$

is attained by the Chebyshev polynomial of first kind with degree n, at points $x=-1$ and $x=1$.

Consider the change of variable

$$t:=\frac{2}{b-a}x-\frac{b+a}{b-a}.$$

When $x=b$, it holds

$$t=\frac{2}{b-a}b-\frac{b+a}{b-a}=\frac{2b-b-a}{b-a}=1$$

and when $x=a$, $t=-1$.

Then,

$$\pm W\cos\left(n\arccos t\right)=W\,T_n\left(t\right)$$

where, T_n is the Chebyshev polynomial of first kind with degree n.

Furthermore, the i-th derivative of the polynomial [A3.1] at point $t=1$ is

$$\left(\pm W\cos\left(n\arccos t\right)\right)_{t=1}^{(i)}=\pm W\left(\frac{2}{b-a}\right)^i T_n^{(i)}\left(1\right). \qquad \text{[A3.2]}$$

The polynomial T_n satisfies the recursive relation [A2.1], hence

$$\left(1-t^2\right)T_n''\left(t\right)-tT_n'\left(t\right)+n^2T_n\left(t\right)=0.$$

Derivating $k-1$ times with respect to t we get

$$\left(1-t^2\right)T_n^{(k+1)}\left(t\right)-\left(2i-1\right)tT_n^{(k)}\left(t\right)+\left[n^2-(i-1)^2\right]T_n^{(k-1)}\left(t\right)=0.$$

With $t=1$, and, in turn, $k=1,...,i$,

$$T_n'\left(1\right)=n^2T_n\left(1\right)=n^2,\ 3T_n''\left(1\right)=\left(n^2-1^2\right)T_n'\left(1\right),...,$$

$$\left(2i-1\right)T_n^{(i)}\left(1\right)=\left[n^2-(i-1)^2\right]T_n^{(i-1)}\left(1\right)$$

which yields

$$T_n^{(i)}(1) = \frac{n^2(n^2-1)\dots\left(n^2-(i-1)^2\right)}{(2i-1)!!}. \qquad \text{[A3.3]}$$

From [A3.2] and [A3.3], we obtain

$$\left|(\pm W\cos(n\arccos t))_{t=1}^{(i)}\right|$$
$$= W\left(\frac{2}{b-a}\right)^i \frac{n^2(n^2-1)\dots\left(n^2 \quad (i \quad 1)^2\right)}{(2i-1)!!}.$$

The same arguments apply with $x = a$. □

We now consider the asymptotic proof by Bernstein (see [VIT 51]).

THEOREM A3.4 (Bernstein).– Let $P(x) := \sum_{j=0}^{n} a_j x^j$ be a polynomial with real coefficients and degree n. If $\sup_{x\in[a,b]}|P(x)| \le W$, then for all x in (a,b) and for all i in $\mathbb{N}$, it holds

$$\left|P^{(i)}(x)\right| \le \frac{n^2(n^2-1)\dots\left(n^2-(i-1)^2\right)}{(2i-1)!!}\left(\frac{2}{b-a}\right)^i W\left(1+\mathcal{O}\left(\frac{1}{n^2}\right)\right).$$

PROOF.– Let $P^{(i)}$ be a polynomial with degree n, defined on $[a,b]$ such that

$$|P(x)| \le W,$$

for $x \in [a,b]$.

Let

$$M := \frac{n^2(n^2-1)\dots\left(n^2-(i-1)^2\right)}{(2i-1)!!}\left(\frac{2}{b-a}\right)^i W,$$

We then have for $x = a$

$$\left|\left(P^{(i)}(x)\right)_{x=a}\right| \le M,$$

and in $x = b$

$$\left|\left(P^{(i)}(x)\right)_{x=b}\right| \le M.$$

By contradiction, assume that for any polynomial with degree n

$$\left|\left(P^{(i)}(x)\right)_{x=b}\right| > M.$$

Consider the polynomial

$$Q(x) := P(x) - \frac{\left(P^{(i)}(x)\right)_{x=b}}{M} W\, T_n(x),$$

where, as above

$$T_n(x) := \cos\left(n \arccos \frac{2x-a-b}{b-a}\right) = \cos(n \arccos x).$$

The polynomial Q alternates sign $n+1$ times at points

$$x_i := \frac{b+a}{2} + \frac{b-a}{2} \cos i\frac{\pi}{n},$$

$i = 0, ..., l.$

Any of the x_i s is a root of T_n; therefore, T_n has n roots inside $[a,b]$ and a root at $x = x_0 = b$. Now $|T_n(x)| \leq 1$, being a Chebyshev polynomial. Also by hypothesis

$$\left|\frac{\left(P^{(i)}(x)\right)_{x=b}}{M}\right| > 1$$

and

$$|P(x)| < W.$$

If x is a point where $T_n = 1$ or $T_n = -1$, it holds either

$$P(x) - \frac{\left(P^{(i)}(x)\right)_{s=\bar{s}}}{M} W\, T_n(x) \geq 0,$$

or

$$P(x) - \frac{\left(P^{(i)}(x)\right)_{x=b}}{M} W\, T_n(x) \leq 0.$$

It follows that the polynomial Q with degree n assumes negative and positive alternating values on the $n+1$ points x_i, and therefore, has n roots inside $[a,b]$.

Consequently its i-th derivative $Q^{(i)}$ has $n-i$ roots inside $[a,b]$. Furthermore, the mapping

$$x \to Q^{(i)}(x) := P^{(i)}(x) - \frac{\left(P^{(i)}(x)\right)_{x=b}}{M} W\, T_n^{(i)}(x)$$

takes value 0 in $x=b$. For $T_n^{(i)}$, it holds

$$WT_n^{(i)}(b) = M$$

by the above partial proof of the Markov theorem A3.3. Hence,

$$\begin{aligned} Q^{(i)}(b) &:= P^{(i)}(b) - \frac{\left(P^{(i)}(x)\right)_{s=\overline{s}}}{M} W\, T_n^{(i)}(b) \\ &= P^{(i)}(b) - P^{(i)}(b) = 0. \end{aligned}$$

It follows that $Q^{(i)}$ with degree $n-i$ has $n-i+1$ roots. This implies that $Q^{(i)}$ is the null polynomial, and therefore, Q has degree at most $i-1$. Since Q has $n > i-1$ roots, it follows that Q is the null polynomial. This implies that

$$\left|\left(P^{(i)}(x)\right)_{x=b}\right| > M$$

cannot hold.

The same arguments hold if we suppose that

$$\left|\left(P^{(i)}(x)\right)_{x=a}\right| > M.$$

Therefore, we have proved that

$$\left|\left(P^{(i)}(x)\right)_{x=b}\right| \leq M \text{ and } \left|\left(P^{(i)}(x)\right)_{x=a}\right| \leq M.$$

Let $x \in [a,b]$ and consider the two following subintervals (a,x), (x,b). Apply,

$$\left|\left(P^{(i)}(x)\right)_{x=a}\right| \leq M,$$

in (a,x) to obtain

$$\left|P^{(i)}(x)\right| \leq \frac{n^2\left(n^2-1\right)\ldots\left(n^2-(i-1)^2\right)}{(2i-1)!!}\left(\frac{2}{x-a}\right)^i W.$$

In the same way apply

$$\left|\left(P^{(i)}(x)\right)_{x=b}\right| \leq M,$$

to the interval (x, b), to get

$$\left|P^{(i)}(x)\right| \leq \frac{n^2\left(n^2-1\right)...\left(n^2-(i-1)^2\right)}{(2i-1)!!}\left(\frac{2}{b-x}\right)^i W.$$

Make the following change of variable

$$\varphi :]a,b[\ \rightarrow \ ]-1,1[\ ; x \ \mapsto \ t = \varphi(x) := \frac{2x-a-b}{b-a}.$$

This yields

$$\left|P^{(i)}(x)\right| \leq \frac{n^2\left(n^2-1\right)...\left(n^2-(i-1)^2\right)}{(2i-1)!!}\left(\frac{2}{1+|x|}\right)^i W.$$

Define,

$$\Psi(x) := \frac{n^2\left(n^2-1\right)...\left(n^2-(i-1)^2\right)}{(2i-1)!!}\left(\frac{2}{1+|x|}\right)^i W.$$

The function Ψ is a decreasing function of $|x|$. By Corollary A3.2, we have

$$\left|\frac{d^i P(x)}{dx^i}\right| \leq \left(\frac{i}{(b-x)(x-a)}\right)^{\frac{i}{2}} n(n-1)...(n-i+1)L.$$

Using the above change of variable φ,

$$\left|\frac{d^i P(x)}{dx^i}\right| \leq \Psi_1(x)$$

where

$$\Psi_1(x) := \left(\frac{i}{(1-|x|)(1+|a|)}\right)^{\frac{i}{2}} n(n-1)...(n-i+1)L.$$

The function Ψ_1 is an increasing function of $|x|$. Henceforth, the upper limit of $\left|\frac{d^k P(x)}{dx^k}\right|$ holds when $|x|$ equals the positive root of the equation

$$\Psi(x) = \Psi_1(x),$$

i.e.

$$\frac{1+|x|}{1-|x|} = \frac{4}{i}\left(\frac{n(n+1)\dots(n+i-1)}{(2i-1)!!}\right)^{\frac{2}{i}},$$

from which

$$\frac{2}{1+|a|} = 1 + \frac{i}{4}\left(\frac{(2i-1)!!}{n(n+1)\dots(n+i-1)}\right)^{\frac{2}{i}},$$

and, since $\left|P^{(i)}(x)\right| \le \Psi(x)$

$$\left|\frac{d^i P(x)}{dx^i}\right| \le (1+\varepsilon_n)\frac{n^2(n^2-1)\dots\left(n^2-(i-1)^2\right)}{(2i-1)!!}L,$$

with

$$1+\varepsilon_n := \left(1 + \frac{i}{4}\left(\frac{(2i-1)!!}{n(n+1)\dots(n+i-1)}\right)^{\frac{2}{i}}\right)^i.$$

For fixed i

$$\begin{aligned} 1+\varepsilon_n &= \left(1 + \frac{i}{4n^2}\left(\frac{(2i-1)!!}{(1+\frac{1}{n})\dots(1+\frac{i-1}{n})}\right)^{\frac{2}{i}}\right)^i \\ &= 1 + \mathcal{O}\left(\frac{1}{n^2}\right). \end{aligned}$$

□

Bibliography

[ACH 92] ACHIESER N.I., *Theory of Approximation*, Dover Publications Inc., New York, 1992.

[AND 62] ANDERSON T.W., "The choice of the degree of a polynomial regression as a multiple decision problem", *Ann. Math. Statist.*, vol. 33, pp. 255–265, 1962.

[ATK 07] ATKINSON A.C., DONEV A.N., TOBIAS R.D., *Optimum Experimental Designs, with SAS*, Oxford University Press, Oxford, 2007.

[BAR 12] BARTROFF J., "A new characterization of Elfving's method for high dimensional computation", *J. Statist. Plann. Inference*, vol. 142, no. 4, pp. 863–871, 2012.

[BAZ 06] BAZARAA M.S., SHERALI H.D., SHETTY C.M., *Nonlinear Programming*, 3rd ed., John Wiley & Sons, Hoboken, NJ, 2006.

[BER 18] BERNSTEIN S., "Quelques remarques sur l'interpolation", *Math. Ann.*, vol. 79, nos. 1–2, pp. 1–12, 1918.

[BER 05] BERGER M.P.F., WONG W. (eds.), *Applied Optimal Designs*, John Wiley & Sons, Chichester, 2005.

[BOG 06] BOGLE I. D.L., ŽILINSKAS J. (eds.), *Computer Aided Methods in Optimal Design and Operations*, World Scientific Publishing Co., 2006.

[BOO 82] DE BOOR C., RICE J.R., "Extremal polynomials with application to Richardson iteration for indefinite linear systems", *SIAM J. Sci. Statist. Comput.*, vol. 3, no. 1, pp. 47–57, 1982.

[BOX 59] BOX G.E.P., LUCAS H.L., "Design of experiments in non-linear situations", *Biometrika*, vol. 46, no. 1/2, pp. 77–90, 1959.

[BRA 98] BRATTI G., *Breviario di Analisi Matematica*, Librenia Progetto Padova, 1998.

[BRO 07] BRONIATOWSKI M., CELANT G., "Optimality and bias of some interpolation and extrapolation designs", *J. Statist. Plann. Inference*, vol. 137, no. 3, pp. 858–868, 2007.

[BRO 12] BRONIATOWSKI M., CELANT G., DI BATTISTA M. *et al.*, "Upper bounds for the error in some interpolation and extrapolation designs", *Comm. Statist. Theory Methods*, vol. 41, nos. 16–17, pp. 3002–3019, 2012.

[BRO 14] BRONIATOWSKI M., CELANT G., "Some overview on unbiased interpolation and extrapolation designs", *Publ. Inst. Stat.Univ. Paris*, vol. 58, no. 3, pp. 27–60, 2014.

[BRU 97] BRUTMAN L., "Lebesgue functions for polynomial interpolation—a survey", *Ann. Numer. Math.*, vol. 4, nos. 1–4, pp. 111–127, 1997.

[BRY 08] BRYAN B.Y.M., *Linear Functional Analysis*, Springer, 2008.

[CEL 03] CELANT G., "Extrapolation and optimal designs for accelerated runs", *Ann. I.S.U.P.*, vol. 47, no. 3, pp. 51–84, 2003.

[CEL 14] CELA A., BEN GAID M., LI X.-G. *et al.*, *Optimal Design of Distributed Control and Embedded Systems*, Springer, 2014.

[CHE 53] CHERNOFF H., "Locally optimal designs for estimating parameters", *Ann. Math. Statistics*, vol. 24, pp. 586–602, 1953.

[CHE 99] CHERNOFF H., "Gustav Elfving's impact on experimental design", *Statist. Sci.*, vol. 14, no. 2, pp. 201–205, 1999.

[COA 66] COATMÉLEC C., "Approximation et interpolation des fonctions différentiables de plusieurs variables", *Ann. Sci. École Norm. Sup.*, vol. 83, pp. 271–341, 1966.

[DET 93] DETTE H., "Bayesian D-optimal and model robust designs in linear regression models", *Statistics*, vol. 25, no. 1, pp. 27–46, 1993.

[DET 95] DETTE H., STUDDEN W., "Optimal designs for polynomial regression when the degree is not known", *Statist. Sinica*, vol. 5, no. 2, pp. 459–473, 1995.

[DET 96] DETTE H., WONG W.K., "Robust optimal extrapolation designs", *Biometrika*, vol. 83, no. 3, pp. 667–680, 1996.

[DET 09] DETTE H., HOLLAND-LETZ T., "A geometric characterization of c-optimal designs for heteroscedastic regression", *Ann. Statist.*, vol. 37, no. 6B, pp. 4088–4103, 2009.

[DOW 54] DOWNTON F., "Least-squares estimates using ordered observations", *Ann. Math. Statistics*, vol. 25, pp. 303–316, 1954.

[DZY 08] DZYADYK V.K., SHEVCHUK I.A., *Theory of Uniform Approximation of Functions by Polynomials*, Walter de Gruyter GmbH & Co. KG, Berlin, 2008.

[EHL 65] EHLICH H., ZELLER K., "Numerische Abschätzung von Polynomen", *Z. Angew. Math. Mech.*, vol. 45, pp. T20–T22, 1965.

[EHR 55] EHRENFELD S., "On the efficiency of experimental designs", *Ann. Math. Statist.*, vol. 26, pp. 247–255, 1955.

[ELF 52] ELFVING G., "Optimum allocation in linear regression theory", *Ann. Math. Statistics*, vol. 23, pp. 255–262, 1952.

[ELF 54] ELFVING G., "Geometric allocation theory", *Skand. Aktuarietidskr.*, vol. 37, pp. 170–190, 1954.

[ERD 43] ERDOS P., "On some convergence properties of the interpolation polynomials", *Ann. of Math.*, vol. 44, pp. 330–337, 1943.

[ERD 58] ERDOS P., "Problems and results on the theory of interpolation. I", *Acta Math. Acad. Sci. Hungarly*, vol. 9, pp. 381–388, 1958.

[ERD 61] ERDOS P., "Problems and results on the theory of interpolation. II", *Acta Math. Acad. Sci. Hungary*, vol. 12, pp. 235–244, 1961.

[ERD 80] ERDOS P., VÉRTESI P., "On the almost everywhere divergence of Lagrange interpolatory polynomials for arbitrary system of nodes", *Acta Math. Acad. Sci. Hungary*, vol. 36, nos. 1–2, pp. 71–89, 1980.

[ERD 81] ERDOS P., VÉRTESI P., "Correction of some misprints in our paper: "On almost everywhere divergence of Lagrange interpolatory polynomials for arbitrary system of nodes"", *Acta Math. Acad. Sci. Hungary*, vol. 38, nos. 1–4, p. 263, 1981.

[FAB 14] FABER G., "Uber die interpolatorische Darstellung stetiger Funktionen", *Jahresber. Deutt. Math. Verein*, pp. 192–210, 1914.

[FAN 99a] FANG Z., Robust extrapolation designs for linear models, Thesis, University of Alberts, Canada, 1999.

[FAN 99b] FANG Z., WIENS D.P., "Robust extrapolation designs and weights for biased regression models with heteroscedastic errors", *Canad. J. Statist.*, vol. 27, no. 4, pp. 751–770, 1999.

[FAN 00] FANG Z., "Robust extrapolation designs for biased polynomial models", *J. Statist. Plann. Inference*, vol. 87, no. 1, pp. 135–147, 2000.

[FAN 03a] FAN S.K., CHALONER K., "A geometric method for singular c-optimal designs", *J. Statist. Plann. Inference*, vol. 113, no. 1, pp. 249–257, 2003.

[FAN 03b] FANG Z., "Extrapolation designs with constraints", *Canad. J. Statist.*, vol. 31, no. 4, pp. 457–468, 2003.

[FED 72] FEDOROV V.V., MALYUTOV M.B., "Optimal designs in regression problems", *Math. Operationsforsch. Statist.*, vol. 3, no. 4, pp. 281–308, 1972.

[FEL 99] FELLMAN J., "Gustav Elfving's contribution to the emergence of the optimal experimental design theory", *Statist. Sci.*, vol. 14, no. 2, pp. 197–200, 1999.

[FID 04] FIDALGO L., DIAZ R., "Elfving's method for m-dimensional models", *Metrika*, 2004.

[GAL 79] GALIL Z., KIEFER J., "Extrapolation designs and Φ_p-optimum designs for cubic regression on the q-ball", *J. Statist. Plann. Inference*, vol. 3, no. 1, pp. 27–38, 1979.

[GAR 54] DE LA GARZA A., "Spacing of information in polynomial regression", *Ann. Math. Statistics*, vol. 25, pp. 123–130, 1954.

[GER 88] GERONIMO J.S., VAN ASSCHE W., "Orthogonal polynomials on several intervals via a polynomial mapping", *Trans. Amer. Math. Soc.*, vol. 308, no. 2, pp. 559–581, 1988.

[GUE 58] GUEST P.G., "The spacing of observations in polynomial regression", *Ann. Math. Statist.*, vol. 29, pp. 294–299, 1958.

[GÜN 80] GÜNTTNER R., "Evaluation of Lebesgue constants", *SIAM J. Numer. Anal.*, vol. 17, no. 4, pp. 512–520, 1980.

[HER 72] HERZBERG A., COX D.R., "Some optimal designs for interpolation and extrapolation", *Biometrika*, vol. 59, pp. 551–561, 1972.

[HIL 56] HILDEBRAND F.B., *Introduction to Numerical Analysis*, McGraw-Hill, 1956.

[HOE 58] HOEL P., "Efficiency problems in polynomial estimation", *Ann. Math. Statist.*, vol. 29, pp. 1134–1145, 1958.

[HOE 62] HOEL P., "Some properties of optimal spacing in polynomial estimation", *Ann. Inst. Statist. Math.*, vol. 13, pp. 1–8, 1961/1962.

[HOE 64] HOEL P., LEVINE A., "Optimal spacing and weighting in polynomial prediction", *Ann. Math. Statist.*, vol. 35, pp. 1553–1560, 1964.

[HOE 65a] HOEL P., "Minimax designs in two dimensional regression", *Ann. Math. Statist.*, vol. 36, pp. 1097–1106, 1965.

[HOE 65b] HOEL P., "Optimum designs for polynomial extrapolation", *Ann. Math. Statist.*, vol. 36, pp. 1483–1493, 1965.

[HOE 66] HOEL P., "A simple solution for optimal Chebyshev regression extrapolation", *Ann. Math. Statist.*, vol. 37, pp. 720–725, 1966.

[HOE 81] HOEL P., "Regression systems for which optimal extrapolation designs require exactly $k + 1$ points", *Ann. Statist.*, vol. 9, no. 4, pp. 909–912, 1981.

[HUA 88] HUANG M., STUDDEN W.J., "Model robust extrapolation designs", *J. Statist. Plann. Inference*, vol. 18, no. 1, pp. 1–24, 1988.

[ISA 94] ISAACSON E., KELLER H.B., *Analysis of Numerical Methods*, Dover Publications Inc., New York, 1994.

[JOH 82] JOHNSON L.W., RIESS R.D., *Numerical Analysis*, 2nd ed., Addison-Wesley Publishing Co., Reading, MA, 1982.

[KAM 61] KAMMERER W., "Polynomial approximations to finitely oscillating functions", *Math. Comp.*, vol. 15, pp. 115–119, 1961.

[KAR 66] KARLIN S., STUDDEN W.J., *Tchebycheff Systems: Applications in Analysis and Statistics*, John Wiley & Sons, New, 1966.

[KIE 58] KIEFER J., "On the nonrandomized optimality and randomized nonoptimality of symmetrical designs", *Ann. Math. Statist.*, vol. 29, pp. 675–699, 1958.

[KIE 59] KIEFER J., WOLFOWITZ J., "Optimum designs in regression problems", *Ann. Math. Statist.*, vol. 30, pp. 271–294, 1959.

[KIE 60] KIEFER J., WOLFOWITZ J., "The equivalence of two extremum problems", *Canad. J. Math.*, vol. 12, pp. 363–366, 1960.

[KIE 64] KIEFER J., WOLFOWITZ J., "Optimum extrapolation and interpolation designs", *Ann. Inst. Statist. Math.*, vol. 16, pp. 295–303, 1964.

[KIE 65] KIEFER J., WOLFOWITZ J., "On a theorem of Hoel and Levine on extrapolation designs", *Ann. Math. Statist.*, vol. 36, pp. 1627–1655, 1965.

[KIE 74] KIEFER J., "General equivalence theory for optimum designs (approximate theory)", *Ann. Statist.*, vol. 2, pp. 849–879, 1974.

[KOL 74] KOLMOGOROV A., FOMINE S., TIHOMIROV V.M., *Eléments de la théorie des fonctions et de l'analyse fonctionnelle*, Mir Publishers, Moscow, 1974.

[KOL 81] KOLMOGOROV A.N., FOMIN S.V., *Elementy teorii funktsii i funktsionalnogo analiza*, Nauka, Moscow, 1981.

[LEB 69] LEBEDEV V.I., "Iteration methods for the solution of operator equations with a spectrum that lies on several intervals", *USSR Compt. Math. and Math Phys.*, vol 9, pp. 17–24, 1969.

[LEF 04] LEFÈVRE V., Arithmétique flottante, INRIA, Report no. RR-5105, 2004.

[LEV 66] LEVINE A., "A problem in minimax variance polynomial extrapolation", *Ann. Math. Statist.*, vol. 37, pp. 898–903, 1966.

[LLO 52] LLOYD E.H., "On the estimation of variance and covariance", *Proc. Roy. Soc. Edinburgh. Sect. A.*, vol. 63, pp. 280–289, 1952.

[LUT 65] LUTTMANN F.W., RIVLIN T.J., "Some numerical experiments in the theory of polynomial interpolation", *IBM J. Res. Develop.*, vol. 9, pp. 187–191, 1965.

[MEL 06] MELAS V., *Functional Approach to Optimal Experimental Design*, Springer, New York, 2006.

[MIT 74] MITCHELL T., "An algorithm for the construction of D-optimal experimental designs", *Technometrics*, vol. 16, pp. 203–210, 1974.

[MÜL 97] MÜLLER C.H., *Robust Planning and Analysis of Experiments*, Springer-Verlag, New York, 1997.

[NOR 99] NORDSTROM K., "The life and work of Gustav Elfving", *Statist. Sci.*, vol. 14, no. 2, pp. 174–196, 1999.

[PÁZ 86] PÁZMAN A., *Foundations of Optimum Experimental Design*, D. Reidel Publishing Co., Dordrecht, 1986.

[PEH 99] PEHERSTORFER F., SCHIEFERMAYR K., "Description of extremal polynomials on several intervals and their computation", *Acta Math. Hungar.*, vol. 83, nos. 1–2, pp. 27–58, 59–83, 1999.

[PEH 01] PEHERSTORFER F., "Deformation of minimal polynomials and approximation of several intervals by an inverse polynomial mapping", *J. Approx. Theory*, vol. 111, no. 2, pp. 180–195, 2001.

[PEH 03] PEHERSTORFER F., "Zeros of polynomials orthogonal on several intervals", *Int. Math. Res. Not.*, no. 7, pp. 361–385, 2003.

[PEH 09] PEHERSTORFER F., "Extremal problems of Chebyshev type", *Proc. Amer. Math. Soc.*, vol. 137, no. 7, pp. 2351–2361, 2009.

[PEH 11] PEHERSTORFER F., "Orthogonal polynomials on several intervals: accumulation points of recurrence coefficients and of zeros", *J. Approx. Theory*, vol. 163, no. 7, pp. 814–837, 2011.

[PRO 13] PRONZATO L., PÁZMAN A., *Design of Experiments in Nonlinear Models*, Springer, New York, 2013.

[PUK 93] PUKELSHEIM F., *Optimal Design of Experiments*, John Wiley & Sons Inc., New York, 1993.

[QUA 07] QUARTERONI A., SACCO R., SALERI F., *Numerical Mathematics*, 2nd ed., Springer-Verlag, Berlin 2007.

[RIC 64] RICE J.R., *The Approximation of Functions. Vol. I: Linear theory*, Addison-Wesley Publishing Co., Reading, MA, 1964.

[RIV 74] RIVLIN T., *The Chebyshev Polynomials*, John Wiley & Sons, New York, 1974.

[SAN 91] SANSONE G., *Orthogonal Functions*, Dover Publications Inc., New York, 1991.

[SAU 83] SAUER F.W., "Algorithm 604. A FORTRAN program for the calculation of an extremal polynomial", *ACM Trans. Math. Software*, vol. 9, no. 3, pp. 381–383, 1983.

[SCH 96] SCHWABE R., *Optimum Designs for Multi-Factor Models*, Springer-Verlag, New York, 1996.

[SIL 80] SILVEY S.D., *Optimal Design*, Chapman & Hall, London, 1980.

[SMI 18] SMITH K., "On the standard deviations of adjusted and interpolated values of an observed polynomial function and its constants and the guidance they give towards a proper choice of the distribution of observations", *Biometrika*, vol. 12, no. 1/2, pp 1–85, 1918.

[SPR 84] SPRUILL C., "Optimal designs for minimax extrapolation", *J. Multivariate Anal.*, vol. 15, no. 1, pp. 52–62, 1984.

[SPR 85] SPRUILL C., "Model robustness of Hoel-Levine optimal designs", *J. Statist. Plann. Inference*, vol. 11, no. 2, pp. 217–225, 1985.

[SPR 86] SPRUILL C., Optimal Experimental Designs, Technical Report Georgia Institute of Technology, DMS 8401, 1986.

[SPR 87a] SPRUILL C., "Optimal designs for interpolation", *J. Statist. Plann. Inference*, vol. 16, no. 2, pp. 219–229, 1987.

[SPR 87b] SPRUILL C., "Optimal extrapolation of derivatives", *Metrika*, vol. 34, no. 1, pp. 45–60, 1987.

[SPR 90] SPRUILL C., "Good designs for polynomial extrapolation", *J. Statist. Plann. Inference*, vol. 26, no. 2, pp. 149–159, 1990.

[STI 09] STIDHAM JR. S., *Optimal Design of Queueing Systems*, CRC Press, Boca Raton, FL, 2009.

[STU 68] STUDDEN W.J., "Optimal designs on Tchebycheff points", *Ann. Math. Statist*, vol. 39, pp. 1435–1447, 1968.

[STU 71] STUDDEN W.J., "Elfving's theorem and optimal designs for quadratic loss", *Ann. Math. Statist.*, vol. 42, pp. 1613–1621, 1971.

[STU 76] STUDDEN W.J., TSAY J., "Remez's procedure for finding optimal designs", *Ann. Statist.*, vol. 4, no. 6, pp. 1271–1279, 1976.

[SZE 75] SZEGŐ G., *Orthogonal Polynomials*, 4th ed., Colloquium Publications, 1975.

[TIT 73] TITTERINGTON D.M., "A method of extremum adaptation", *J. Inst. Math. Appl.*, vol. 11, pp. 297–315, 1973.

[TOT 11] TOTIK V., "The norm of minimal polynomials on several intervals", *J. Approx. Theory*, vol. 163, no. 6, pp. 738–746, 2011.

[VÉR 90] VÉRTESI P., "Optimal Lebesgue constant for Lagrange interpolation", *SIAM J. Numer. Anal.*, vol. 27, no. 5, pp. 1322–1331, 1990.

[VIT 51] VITALI G., SANSONE G., *Moderna teoria delle funzioni di variabile reale*, 3rd ed., Nicola Zanichelli, Bologna, 1951.

[WAL 43] WALD A., "On the efficient design of statistical investigations", *Ann. Math. Statistics*, vol. 14, pp. 134–140, 1943.

[WHI 73] WHITTLE P., "Some general points in the theory of optimal experimental design", *J. Roy. Statist. Soc. Ser. B*, vol. 35, pp. 123–130, 1973.

[WIE 08] WIENS D.P., XU X., "Robust prediction and extrapolation designs for misspecified generalized linear regression models", *J. Statist. Plann. Inference*, vol. 138, no. 1, pp. 30–46, 2008.

[WU 78] WU C., WYNN H.P., "The convergence of general step-length algorithms for regular optimum design criteria", *Ann. Statist.*, vol. 6, no. 6, pp. 1273–1285, 1978.

[WYN 70] WYNN H.P., "The sequential generation of D-optimum experimental designs", *Ann. Math. Statist.*, vol. 41, pp. 1655–1664, 1970.

[WYN 72] WYNN H.P., "Results in the theory and construction of D-optimum experimental designs", *J. Roy. Statist. Soc. Ser. B*, vol. 34, pp. 133–147, 170–186, 1972.

[WYN 75] WYNN H.P., "Simple conditions for optimum design algorithms", *(Proc. Internat. Sympos., Colorado State Univ., Ft. Collins, Colo.)*, pp. 571–579, North-Holland, Amsterdam, 1975.

[XU 06] XU X., Robust prediction and extrapolation designs, with applications to accelerated life testing, PhD Thesis, University of Alberta, Canada, 2006.

[XU 09] XU X., "Robust prediction and extrapolation designs for censored data", *J. Statist. Plann. Inference*, vol. 139, no. 2, pp. 486–502, 2009.

[XU 14] XU X., CHEN A., "Robust prediction and extrapolation designs for nonlinear regression with imprecision", *Metron*, vol. 72, no. 1, pp. 25–44, 2014.

Index

A, B, C

D, E, F

G, H, I

L, M

N, O, P

R, S

T, V, W

Other titles from

iSTE

in

Mathematics and Statistics

2015

DE SAPORTA Benoîte, DUFOUR François, ZHANG Huilong
Numerical Methods for Simulation and Optimization of Piecewise Deterministic Markov Processes

DEVOLDER Pierre, JANSSEN Jacques, MANCA Raimondo
Basic Stochastic Processes

LE GAT Yves
Recurrent Event Modeling Based on the Yule Process
(Mathematical Models and Methods in Reliability Set – Volume 2)

2014

COOKE Roger M., NIEBOER Daan, MISIEWICZ Jolanta
Fat-tailed Distributions: Data, Diagnostics and Dependence
(Mathematical Models and Methods in Reliability Set – Volume 1)

MACKEVIČIUS Vigirdas
Integral and Measure: From Rather Simple to Rather Complex

PASCHOS Vangelis Th
Combinatorial Optimization – 3-volume series – 2nd edition
Concepts of Combinatorial Optimization / Concepts and Fundamentals – volume 1
Paradigms of Combinatorial Optimization – volume 2
Applications of Combinatorial Optimization – volume 3

2013

COUALLIER Vincent, GERVILLE-RÉACHE Léo, HUBER Catherine, LIMNIOS Nikolaos, MESBAH Mounir
Statistical Models and Methods for Reliability and Survival Analysis

JANSSEN Jacques, MANCA Oronzio, MANCA Raimondo
Applied Diffusion Processes from Engineering to Finance

SERICOLA Bruno
Markov Chains: Theory, Algorithms and Applications

2012

BOSQ Denis
Mathematical Statistics and Stochastic Processes

CHRISTENSEN Karl Bang, KREINER Svend, MESBAH Mounir
Rasch Models in Health

DEVOLDER Pierre, JANSSEN Jacques, MANCA Raimondo
Stochastic Methods for Pension Funds

2011

MACKEVIČIUS Vigirdas
Introduction to Stochastic Analysis: Integrals and Differential Equations

MAHJOUB Ridha
Recent Progress in Combinatorial Optimization – ISCO2010

RAYNAUD Hervé, ARROW Kenneth
Managerial Logic

2010

BAGDONAVIČIUS Vilijandas, KRUOPIS Julius, NIKULIN Mikhail
Nonparametric Tests for Censored Data

BAGDONAVIČIUS Vilijandas, KRUOPIS Julius, NIKULIN Mikhail
Nonparametric Tests for Complete Data

IOSIFESCU Marius *et al.*
Introduction to Stochastic Models

VASSILIOU PCG
Discrete-time Asset Pricing Models in Applied Stochastic Finance

2008

ANISIMOV Vladimir
Switching Processes in Queuing Models

FICHE Georges, HÉBUTERNE Gérard
Mathematics for Engineers

HUBER Catherine, LIMNIOS Nikolaos *et al.*
Mathematical Methods in Survival Analysis, Reliability and Quality of Life

JANSSEN Jacques, MANCA Raimondo, VOLPE Ernesto
Mathematical Finance

2007

HARLAMOV Boris
Continuous Semi-Markov Processes

2006

CLERC Maurice
Particle Swarm Optimization

Printed and bound by CPI Group (UK) Ltd, Croydon, CR0 4YY
16/09/2026
14963139-0001